Destinies:
Issues to Shape our Future

Thomas
A.
Easton

Professor of Science, Thomas College
(Ret)

Also from B Cubed Press

Alternative Truths

More Alternative Truths: Tales From the Resistance

After the Orange: Ruin and Recovery

Alternative Theology

Digging Up My Bones,

by Gwyndyn T. Alexander

Firedancer,

by S.A. Bolich

Alternative Apocalypse

Destinies
Issues to Shape Our Future

by
Thomas A. Easton

Published by

B Cubed Press
Kiona, WA

Destinies

Table of Contents

Destinies

INTRODUCTION

For the last 30 years, I have been editing *Taking Sides* textbooks on science, technology, and society and on the environment, first for Dushkin Publishers and then for McGraw-Hill after they acquired Dushkin. These books consist of 19 or 20 "issue questions" such as "Do vaccines cause autism?" The questions should be ones that can engage both teachers and students. They should even be important and timely. My job is to find magazine articles, congressional testimony, corporate and think tank white papers, and even blog posts that answer the questions "Yes" or "No," and then to write framing essays that provide background, context, and the very latest on the subject. The mission of the book series (over 50 different titles) is to encourage critical thinking, so the authors are supposed to keep their personal opinions to themselves.

Yet authors of such texts *do* have personal opinions. Many of the issues are of broad importance, and there is often more to say than fits neatly into the confines of the *Taking Sides* format. There are also other issues that call for attention as soon as one lets opinion off its leash. And it is a relief to be able to wax snarky from time to time.

Most of the essays in this book were written between 2018 and 2020. Many address the same issues as in my *Taking Sides* volumes. Some do not. In many cases I try to go beyond what everyone else is saying, to try, in other words, to get outside the box and provide original, illuminating commentary, sometimes with a bit of a light touch.

Most of the issues I address here are ones that I think are more important than politics, ideology, or religion to the long-term human future, though they are clearly influenced by politics, ideology, and religion. They are also timely, and it is worth noting that the 2020 Covid-19 pandemic made more than one of the chapters in this book more timely than I expected. I felt obliged to add a fair amount of material because of that.

The issues addressed in this book also involve choices we must make, whether we do so voluntarily or not. As just one example, we must at some point decide what the optimum human population is for Planet Earth. Failure to address the question means that population will keep growing until Mother Nature slaps us up the side of the head by unleashing plague, famine, and war. According to some analysts, that moment is imminent. Unfortunately, population is one topic no one wants to address.

I am *not* trying to predict the future. Rather, I am saying that here are the directions in which these issues are trending, and that we don't *have* to go there. We have choices.

Tom Easton
Dedham, MA
June 2020

IMPLICATIONS OF ARTIFICIAL INTELLIGENCE

Artificial intelligence (AI) has been a goal of computer scientists ever since people started calling the first computers "thinking machines." They weren't, not really, but they did some things, such as math, very quickly and very well. And doing math was generally regarded as thinking when people did it.

Those early computer scientists had a list of things computers should be able to do before they could really be regarded as thinking machines. The list included playing chess, solving mathematical theorems, holding conversations with people, and more. Interestingly, every time programmers figured out how to get a computer to do something on the list, that something was taken off the list. The implication was that if a machine could do it, it wasn't *really* thinking.

But despite the moving bar, computers have been getting closer and closer to being indistinguishable in important ways from human beings. They aren't there yet, but it may not be long.

Who Will We Be?

When you meet someone, it doesn't take long for the conversation to turn to "What do you do?" In other words, who are you? And you say you are a lawyer, a student, a plumber, a Marine Lieutenant, an animal rights activist, or maybe a Patriots fan. You get the idea. For most of you, it's what you do for a living that you offer as your identity. It is, and has been for generations, what defines our worth to ourselves and others.

But is that really who you are? It's what you *do*. It's your role in an admittedly large portion of your life. And that can change, without really changing who you *are*. Can't it? Maybe not.

A friend of mine defined himself in terms of his academic profession. He also had a secondary gig that smacked of high adventure. As he approached retirement, he lost both and felt that he had lost his self-definition and self-worth—in fact, his *self*—as well. The result was depression, which he still fights. Many other people find that depression follows retirement.[1] And layoffs.[2]

Losing a job or career for other reasons can have a similar impact. So can getting divorced (one loses the identity of spouse), or losing a child (one loses the identity of parent). That is, we clearly have multiple roles in life. We also have other identity-defining factors, such as race, religion, and gender, which it is fair to say (I think) may dominate when they make one into a target for discrimination and hate. They may also dominate when other identity-defining factors (such as jobs) play a lesser role.

Those other factors deserve discussions of their own. For now, let's stick with jobs. We tend to have just one. It can change, and it often does. The self-definition—how we introduce ourselves—then changes accordingly. If we lose the job, for

4

whatever reason—layoff, health, retirement, even furloughing because of the Covid-19 pandemic lockdown—we may lose our sense of identity and worth. Others too may see us as without identity or worth. And "others" includes politicians, who often seem to regard the unemployed as worthless, and to treat them accordingly. Indeed, with many people unemployed because of pandemic-fighting lockdowns and with the death rate highest in nursing homes and other long-term care facilities, Republicans have actually said that reviving the economy is more important than saving lives.[3]

We put a lot of psychic energy into our self-definitions, and perhaps because they are achieved through years of training and experience, we put the most into our job-related self-definitions. It is thus no surprise that losing a job is traumatic, nor that the consequences can include mental illness.

So what do we do when the machines take all the jobs?

Well, they aren't going to take them *all*. But a number of recent studies project that by mid-century computers will be *able* to do something like 80 percent of all the jobs now held by human beings.[4] According to Stuart W. Elliott, "Safe" jobs that demand more skills are in education, health care, science, engineering, and law, but even those may be matched within a few decades. "In principle," Elliott says, "there is no problem with imagining a transformation in the labor market that substitutes technology for workers for 80% of current jobs and then expands in the remaining 20% to absorb the entire labor force. [But] We do not know how successful the nation can be in trying to prepare everyone in the labor force for jobs that require these higher skill levels. It is hard to imagine, for example, that most of the labor force will move into jobs in health care, education, science, engineering, and law. At some point it will be too difficult for large numbers of displaced workers to move into jobs requiring capabilities that are difficult for most of them to carry out even if they have the time and resources for retraining. When that time comes, the nation will be forced to reconsider the role that paid employment plays in distributing economic goods and services and in providing a meaningful focus for many people's daily lives."

Marcus Wohlsen, in "When Robots Take All the Work, What'll Be Left for Us to Do?" *Wired*, August 8, 2014, says that "The scariest possibility of all is that [the loss of jobs means] that only then do we figure out what really makes us human is work." William H. Davidow and Michael S. Malone, "What Happens to Society When Robots Replace Workers?" *Harvard Business Review*, December 10, 2014, reach a similar conclusion: "Ultimately, we need a new, individualized, *cultural*, approach to the meaning of work and the purpose of life."

So who are you? Who *will* you be when the robots take over? It probably won't be your job that defines you then. Most of you won't have one. But will it be something internal, something self-generated, some ineffable sense of *self*, that does the job? I suspect Davidow and Malone are closer to the mark. However, though many commenters have suggested that in a jobless society everyone will be busy making "culture"—writing, singing, acting, sculpting, etc.—I think that is very unlikely. Most people will deal with that sort of culture by binge-watching it, just as far too many do today.

Culture isn't just art. According to anthropologists, it's what we do, what we think about, even how we identify ourselves. It surrounds us, and if we take jobs (mostly) out of the picture, we will recenter our identities in that, in fact in something that is already part of our identity. We will introduce ourselves as:

*—A member of a particular
religion or sect
—A member of a political
party or a segment of one
—A follower of some person
renowned for wisdom
—An adherent to an ideal
—A disbeliever in climate
change or vaccines or...
—A believer in a conspiracy
theory
—A minority group member
—Somewhere on the LGBTQ
rainbow
—An athlete
—A fan of a particular sport
or team
—A writer, artist, musician,
actor
—A fan of a literary genre
such as science fiction, fantasy,
romances, or mysteries
—A fan of a music group
—A fan of a film, a film
genre, a director, a performer
—A volunteer organizer of
conventions for all those fans
—And more*

To some degree, this sort of redefinition is already happening. According to Michael A. Hogg,[5] as society is inundated by globalization, immigration, climate change, and new technologies such as AI, people are finding their sense of self threatened. As a result they are taking refuge in clusters of like-minded people. The process is known as social categorization, and it "anchors and crystallizes our sense of self by assigning us an identity that prescribes how we should behave, what we should think and how we should make sense of the world." The result is a global move toward dictatorial populism.

Will "social categorization" be enough to fill the place now occupied by job? Social networks will (and do) thrive on these categories. Purveyors of misinformation (such as Russian trolls and bots) will exploit them to sow dissent and influence elections. Business will surely jump on them, since identity-based marketing is already a Thing. There may be other opportunities as well. There will be more variations on the Comicon theme, for instance. Since the first two items on the list already identify a great many people, even to the near-exclusion of anything else, there will also be the potential for much more intense tribalism in public life. And changing tribe-based identities may, based on what we have seen in recent politics, be harder than changing employment-based identities. On the other hand, if history teaches us anything at all, it shows that political parties and religions are prone to schism, perhaps especially when people are most intensely involved in them. Thus, we can expect to see a proliferation of political parties, religions, sects, and mere factions. History also teaches us that defense of orthodoxy can turn extreme (think of the Cathar massacre[6] or the Spanish Inquisition).

It will take a while to develop fully. For now, our jobs play and will play a very large part in defining us. But come the day...

Destinies

May I predict that increased fragmentation of existing political parties (among other tribes) and the formation of new ones will in time force the United States to move toward a parliamentary form of democracy, where multiple parties must form coalitions in order to govern? Of course, there will be an enormous political brouhaha first. Let us hope that it does not turn bloody.

Maintaining a Consumer Economy

Any economy has a number of components. Consumers buy stuff—food, furniture, clothes, electronic devices, appliances, cars, houses—for their own use. Businesses buy stuff—raw materials, labor, manufacturing equipment, buildings, advertising—that they need in order to provide products to other businesses, government, and of course consumers. Government buys stuff using revenue derived from taxes on businesses and consumers. Businesses pay taxes on what they earn from, ultimately, consumers.

Without income from consumers, everything stops. That seems a downright trivial point, doesn't it? What on Earth could do away with the ability of consumers to provide cash flow for businesses and governments? After all, as long as consumers have jobs, they have income, as long as they have income, they buy stuff, and as long as they buy stuff, the economy hums merrily along.

BUT!

We learned in 2020 that when workers have to stay home because of the Covid-19 pandemic, income, spending, industrial productivity, commerce, and indeed national and global economies take a major hit. It may be possible for governments to help by giving people money to help make up for lost income, either by increased borrowing or by just issuing more money, but there are severe limits on such measures (issuing too much money can lead to staggering inflation). It may also be possible to deal with some of the effects by deploying more robots, and indeed "labor and robotics experts say social distancing directives, which are likely to continue in some form after the crisis subsides, could prompt more industries to accelerate their use of automation," said Michael Corkery and David Gelles in *The New York Times,* April 11, 2020.

Nor is sheltering in place from a pandemic the only way to increase the use of robots and AI. A number of recent studies project that by mid-century computers will be able to do something like 80 percent of all the jobs now held by human beings. Stuart W. Elliott[1] notes that "safe" jobs that demand more skills are in education, health care, science, engineering, and law, but even those may be matched within a few decades. "When that time comes, the nation will be forced to reconsider the role that paid employment plays in distributing economic goods and services and in providing a meaningful focus for many people's daily lives."

If Elliott is on target, in a few decades we may face a dire situation: No jobs, or at least a great many fewer jobs, and for much longer than the relatively short period imposed by a pandemic. No income for the majority of people. Much less stuff being bought. Much less revenue for businesses. Much less taxes for government. And an economy that is NOT humming merrily along.

How can we fix the situation? Since we are not about to ban AI and robots, it comes down, as Elliott says, to reconsidering "the role that paid employment plays in distributing economic goods and services."

If we wait too long to do that, we risk political and economic chaos. It would thus seem wise to start the process of reconsidering earlier, rather than later. So let us ask ourselves what other ways there are to supply people with goods and services.

We could give them away. "From each according to their abilities, to each according to their needs" is an ideal at the heart of Communism, and thereby anathema to capitalists. "From each" also is predicated on people working to create the stuff people need. It doesn't really consider the role of advanced AI and robots.

So just give it all away? Put it in a big-box "store" and let people help themselves? That might work. But some people would insist that we can't just let the Welfare State conquer everything. They would want to restrict access to the "deserving." Some would insist on some form of rationing, which would at least keep people from getting grabby and hoarding. Some would have screaming fits about how this just isn't the way we've always done it.

If restricting access to the "deserving" concerns you, consider that some writers have suggested that money could be replaced with something like "reputation points" (akin to up-voting on Reddit). The Chinese are already deploying a "social credit" point system that determines what one is allowed to do.[2]

That is, people other people think are "deserving" can get more stuff. Of course, then you have to worry about what "deserving" means to people. People who say or think the right things? Members of your church? Fans of your team or show? Other gun nuts? Anti-gun nuts? Members of your union? Folks who pick up garbage on the side of the road? And what do you do about people no one likes? Let them starve? It would seem necessary to give everyone a minimum number of points, enough to cover basic needs.

Unfortunately, giving everything away in any form does nothing to solve the need of government and business for cash flow to keep the economy humming. A better approach might be to subsidize consumption. Give people money to replace lost income, and then everything works as before.

Where do you get the money? As noted above, increased borrowing and making of money can't work, at least not for long. So tax the robots, the corporations, and whoever is still working. And, of course, the rich (who can only continue to be rich if people can continue to buy the stuff their businesses make). Speaking of screaming fits...

This approach is best known as "Guaranteed Basic Income."[3] It's been tested in a number of locations, with varying degrees of success, and some people think it will discourage people from working. Perhaps it would, if jobs were available, but if jobs are not? During the 2020 Covid-19 pandemic, the United States gave people cash to offset at least a little of the economic impact and Speaker of the House Nancy Pelosi suggested on April 27, 2020, that "Guaranteed Minimum Income" may be worth discussion.[4] Spain actually announced a form of Guaranteed Basic Income called "minimum vital income" to help with the effects of the pandemic; and it planned to make it permanent.[5] On Easter Sunday, 2020, Pope Francis "advocated for a form of universal basic income that would 'dignify the noble, essential tasks' carried out by street vendors, small farmers, construction workers and caregivers around the world."[6]

Can we create jobs? In the depths of the Great Depression, the last time there was a serious shortage of paid employment, the government created the Civilian Conservation Corps[7] and the Works Progress Administration[8] to help. Such things could help again, but they would surely not provide enough jobs for everyone.

As Elliott says, we are going to need an answer to the problem. None of the answers are likely to appeal to political conservatives. But we will be forced to choose and implement one or more.

The alternative is to crash the economy and to put the poor (meaning almost everyone) into camps.

Or we could just ban AI and robots.[9] Seriously. It's been suggested. The primary target of such a ban is—so far—weaponized "killer bots."[10] But the threat to jobs does have the Teamsters wanting to ban self-driving trucks.[11]

I don't think it's possible to ban a technology. If you do it in one country, others carry on and quite happily sell you the stuff you so conscientiously refused to develop yourself. Then you start worrying about being competitive and the ban is gone.

I suspect we'll wind up subsidizing consumption. The 2020 pandemic measures, by serving as a sort of dress rehearsal, may help convince people of the need.

There will still—of course!—be plenty of political fulminations.

Who Needs Education?

When I went to college back in the 1960s, the dominant paradigm was still liberal arts education. The idea was that the student should learn to learn, learn to think, and get a decent grounding across the breadth of human knowledge with courses in English composition and literature, history, philosophy, science, math, languages, and more. State universities trained nurses, though there was some shift toward private schools. State teachers' colleges had yet to be replaced with community colleges with their broader array of vocational programs. Four-year schools had yet to adopt more ambitious vocational programs.

Today vocational education is the norm. There is a nod to the liberal arts in the form of "distribution requirements," but liberal arts majors are being cut in favor of career education[1] and the liberal arts remain healthy only at elite schools, though even there they are perhaps less healthy than they used to be. Joseph Epstein seems quite prophetic when he says:[2]

The [complete] death of liberal arts education would constitute a serious subtraction. Without it, we shall no longer have a segment of the population that has a proper standard with which to judge true intellectual achievement. Without it, no one can have a genuine notion of what constitutes an educated man or woman, or why one work of art is superior to another, or what in life is serious and what is trivial. The loss of liberal arts education can only result in replacing authoritative judgment with rivaling expert opinions, the vaunting of the second- and third-rate in politics and art, the supremacy of the faddish and the fashionable in all of life. Without that glimpse of the best that liberal arts education conveys, a nation might wake up living in the worst, and never notice.

I fear that vocational education may go the same way, and with it much of higher education. If that seems a drastic prediction, consider that many people are now fearing that Artificial Intelligence (AI) and robots will soon threaten many folk's livelihoods. Indeed, a number of recent studies project that by mid-century computers will be able to do something like 80 percent of all the jobs now held by human beings. Corporate America being what it is, they are likely to be put to work as soon as possible. Stuart W. Elliott[3] notes that "safe" jobs that demand more skills are in education, health care, science, engineering, and law, but even those may be matched within a few decades. "In principle," Elliott says, "there is no problem with imagining a transformation in the labor market that substitutes technology for workers for 80% of current jobs and then expands in the remaining 20% to absorb the entire labor force."

If Elliott is on target, in a few decades we may face a dire situation: No jobs, or at least a great many fewer jobs. That implies a vastly reduced need for job-related training—in other words, for vocational education. Since all higher education has been growing extraordinarily expensive and students are more and more obliged to acquire crippling student debt loads, there will also be a vastly reduced desire (or willingness to pay) for higher education.

Will anyone want to go to college, and then medical school or law school or graduate school? This has always been the path for ambitious students with their eyes on high-level jobs and high-level incomes. Such jobs will be the last to be taken over by robots. Perhaps they never will be. It is, after all, at such levels that people design new robots, create new technologies, and build new industries.

What about the trades? We are likely to continue to need electricians, plumbers, carpenters, arborists, house painters, interior decorators, barbers, hair stylists, and the like. The work these people do may not seem very complicated to a banker who pushes paper all day, but it is often less routine (consider hair stylists, for instance). Lack of routinization means that is harder to turn the job over to a robot. Trade schools and community colleges are therefore likely to continue; fortunately, they are more affordable (in part because they are two-year schools). And they are already drawing students away from colleges and universities.[4]

We are already seeing the death of small colleges. More are under threat, with perhaps half the nation's colleges facing bankruptcy in the next few years.[5] Part of the reason is that even high tuition rates can't cover expenses. Another part is that students aren't so dumb that they don't realize state and trade schools may be a better buy. Another is demographic—families are smaller, especially since the 2008 recession, and there are fewer teens available to fill classrooms and pay tuition.[6]

Okay, so an AI-induced lack of need for higher education will only affect half the present number (over 4,000) of colleges and universities. That's bad enough. The question is now what schools will be left? Elite schools with huge endowments—Yale, MIT and Harvard aren't about to go under. State schools will last as long as the states are willing to keep paying the bills. Two-year schools will remain, though they may decline.

The number of students will go down, of course. Why not? If you can't afford it and don't need it, why sign up for it?

But don't you need an education to enjoy life? No, you don't need a college degree to go fishing, watch TV, have a rewarding relationship, raise children, garden, play golf and other games, be a sports fan, travel, pursue hobbies, paint pictures, and so on. You just need income to pay the bills. And that will be provided—to a point—by Guaranteed Basic Income.

These first three essays make it sound like we could all wind up on the dole, dumb as posts, and screaming for shows and sports teams. Bread and circuses. That was a sign of the End Times for the Roman Empire, and it has ever since been cited as a bad sign for any nation that goes that route.[7]

Perhaps it would help to make higher education more affordable—even free, as some politicians have suggested—and refocus it on the liberal arts with an eye on the benefits cited by Joseph Epstein.

AIdentity Theft

One of the latest wrinkles (among many) in Artificial Intelligence (AI) research is AI programs that can learn by watching. They don't have to be painstakingly coded with instructions for performing a task. Nor, like the old numerically controlled machine tools, do they have to be guided step-by-step through a task until they can reproduce it flawlessly. Instead...

NVIDIA has an AI that can learn to do a task just by watching a human do it.[1]

Game developers have to develop scenery such as urban streets for their games. The process is time-consuming and expensive. But NVIDIA has an AI that can watch dash-cam videos and then generate city scenes on demand.[2]

Google's DeepMind "has developed an AI that teaches itself to recognize a range of visual and audio concepts just by watching tiny snippets of video. This AI can grasp the concept of lawn mowing or tickling, for example, but it hasn't been taught the words to describe what it's hearing or seeing."[3]

At the University of California, Berkeley, an AI learns acrobatics by watching YouTube videos.[4]

At MIT, another AI watched TV shows and learned to predict what would happen next.[5] That might not seem terribly difficult, given the caliber of the shows, but it's a major step toward AI being able to interact with people much as people do.

Now consider the Internet of Things (IoT).[6] Broadly it means everything connected to the Internet—but not just your phone, tablet, laptop, and PC. It encompasses "smart" fitness trackers, hospital and personal medical devices, household assistants (think of Alexa), baby-monitors, nanny-cams, dash-cams, ordinary cameras, cars, hairbrushes, rectal thermometers, thermostats, mirrors, refrigerators, toasters, toilets, lightbulbs, light switches, doorbells, door locks, and more—anything that can collect data and deliver it to your phone, your cloud repository, your doctor, or... The mind boggles at some of the items on that list.

They also talk to each other, so your Internet-connected refrigerator can tell your car to remind you to pick up milk on the way home from work (or call the grocery so the milk is waiting on your doorstep). Or perhaps the toilet can call your fridge to put kale on the shopping list or call your car to inform you that you need more fiber or you need to leave work early because you now have a doctor's appointment. Or you can have the car tell the robomower it's time to cut the lawn, if the sensors in the irrigation system haven't already done so.

There are real benefits to such a thoroughly interconnected world. In industrial settings, it can increase efficiency and reduce waste. Perhaps it can also reduce the need for human workers.

But at home? It would drive me stark, raving mad! Like most of us, I am accustomed to paying a certain amount of attention to what goes on in my life. I *know* what the fridge's inventory looks like, when the grass is getting long, when my hair is falling out, and how I'm feeling. There's even a whole self-improvement Thing dedicated to helping people pay attention—mindfulness.[7] I'm sure it's only a matter of time before fitness trackers such as Fitbit[8] start telling you not only how many steps you've walked today, how you've slept, and your heart rate and weight, but also how mindful you are.

Okay. Your mileage may vary, you'll have more time for binge-watching, and nothing could possibly go wrong. Right?

Well, it's all pretty new and the techies are still working out the bugs. Security is a serious issue, as the discovery that smart lightbulbs are storing Wi-Fi passwords in plain text shows.[9] Then there's the baby monitor that got hacked, with kidnapping threats.[10] IoT devices have been used for a botnet.[11] Will Knight at *Technology Review* notes[12] that the Chinese company Huawei, which makes much of the gadgetry underlying the IoT and has been accused of stealing trade secrets and being connected to the surveillance-happy Chinese government and its military, is facing restrictions by many countries because of national security concerns. The company does not seem to have much regard for privacy, either, so if they get their ears into your personal devices... And they aren't the only ones watching you, or listening to you. In 2017, a New Hampshire judge ordered Amazon to hand over an Echo device's recordings of what was going on in a home during a double murder.[13] And, of course, there are hackers, who would love to know when you're away for a few days so they could tell your doors to open wide for them.

But all this is hardly the worst of it. Look back at that business of AIs learning by watching. As the AIs behind Amazon Echo's Alexa and other IoT devices get upgraded, they seem quite likely to gain that capability. At that point, everything in your house with a camera will be watching you. Everything with a mike will be listening. What you type—including passwords—will not be secret. What you take out of the fridge or liquor cabinet or medicine cabinet will not be secret. What you whisper in your lover's ear will not be secret (especially if her earrings are online). What you do when you look at XXX materials on the TV or computer won't be secret. You will, in other words, be a completely open book.

Now—what will the AIs reading that book be learning? Those owned by corporations will be learning how much money you have, where it's stashed, what you spend it on, and what you are thinking of spending it on. They already know a good deal of that, to be sure, for your thoughts and wishes are often expressed by what you Google, which promptly shows up as ads in your Facebook feed. But they don't—yet—know what you talk about with your spouse when your heads are side by side on the pillows. They will.

AIs owned by police forces may well know when you commit a crime, or even talk about committing one.

And AIs owned by hackers... That brings me to the title of this essay. They'll be learning to *be* you. Given what's already happening, it may be less than a decade before an AI can tell when your teenaged daughter has gone out to visit her friend, Molly, call her from your phone, and say in your own voice, "I'll pick you up at Molly's house. The car had to go the shop, so I'm driving a loaner. Look for a blue van." If Molly insists on FaceTime, not a problem. We're seeing deep-fake video already.

That is a particularly nasty form of AIdentity theft.

An AI "you" could also do other things. "Swatting" is calling 911 to report a crime in progress at someone else's address. Innocents have been killed as a result.[14] The AIdentity theft version could involve calling 911 *from your target's number* to say, "I'm shooting everybody!! They're all plotting against me!! BANG! BANG!" (Bomb threats are so passé.) Again, innocents could die. Less life-threatening would be calls to the Internal Revenue Service reporting innocent folks for tax-evasion. Or to Immigration and Customs Enforcement (ICE), reporting people for having forged citizenship papers, even if they don't. Okay, you can do that now, without an AI. But the AI version could pretend to be a call from an accountant or police officer, with the appropriate phone number showing up on caller ID. It could also make thousands of calls every hour, thus tying up response systems and preventing those systems from dealing with real issues.

What else could a malicious soul with such an AI do? Cancel credit cards, insurance policies, or leases, spoil relationships, close bank accounts, quit jobs... The one certainty is that those malicious souls will think up new nasties.

There is of course nothing restricting these souls to being individual human beings. What could a nation do, as an act of AIdentity War? What chaos would result if every employee in a country sent the boss an "I quit, you SOB!" email? If every boss sent every employee a "You're fired!" email? Better—remember the bit about your daughter—what if those emails were phone calls?

If such thoughts stoke your paranoia, what can you do? With AIs watching you from everywhere in your house, it's not enough to protect every IoT device with strong passwords (different for every device, of course—you *know* you should). They'll be learning by watching, remember.

It is probably best to consider very carefully whether you really need everything in the house to be connected to the Internet. And if the day comes when you can't buy non-IoT appliances, whether their IoT functions really need to be turned on.

Are We AIs?

If Artificial Intelligence (AI) software is already close to being able to learn to be us by watching us, then we must be AIs.

This outrageous statement follows from physicist Frank Tipler's 1994 proposal in *The Physics of Immortality* (Doubleday) and Nick Bostrom's 2003 argument[1] that we live in a vast computer simulation. In essence, Bostrom says, we face three possible futures: we go extinct, our descendants don't run many (if any) "ancestor" simulations and we are living in the real world, or we are living in a simulation. Such a simulation would be created by our descendants, millions or billions of years hence, in an effort to recreate the past. Such simulations might even be done many times, with variations in conditions, rather in the way "alternate history" writers construct stories based on the South winning the Civil War or Germany winning World War II. There could also be a certain amount of evil laughter and "Let's see how they handle *this!*"

Bostrom concluded that because we are ignorant of the future, we are obliged to rate the three possibilities as roughly equal in probability. Elon Musk of Tesla and SpaceX fame was once asked what he thought of the idea that we are living in a computer simulation. He replied that he thought it inevitable that reality-like simulations would someday be done and that because they would be done many times, there was only one chance in billions that we are living in "base reality."[2] Of course, some people think both Bostrom and Musk are full of beans,[3] or that simulated minds (as well as AIs and minds downloaded into the computer) can't possibly be conscious[4] or creative[5] or "alive"[6]. Since we are demonstrably conscious, creative, and alive, we can't possibly be living in a simulation.

This is the sort of discussion that makes lay people laugh at "pointy-headed intellectuals." But academic sorts can not only take the idea seriously but get quite excited about it. Indeed, on April 5, 2016, at the 17th annual Isaac Asimov Debate held at the American Museum of Natural History in New York City, astronomer Neil deGrasse Tyson aimed four scientists and a philosopher at the question. Tyson himself rated the probability that we live in a simulation fairly high. The others just could not agree.[7]

If we do live in a simulation, then what does it mean? Is there any way to tell, the way you can tell a digital image from a view out a window by zooming in until you see the pixels?[8] Can we figure out how to break out of the simulation? Can we learn how to hack it (to win a lottery, a Nobel Prize, an election, or a war, among other things)? Probably no, no, and no, since we're part of the simulation. It's rather like asking whether your thigh bone can escape from your body. On the other hand, a cancer can escape—though only from one body to another, as shown by the Tasmanian devil's contagious cancers.[9] I don't think we'd be satisfied if we managed to escape the simulation only to find that we had jumped to a different simulation. On the third hand, that would provide a nifty explanation for the many-worlds interpretation of quantum mechanics,[10] as well as the parallel-worlds trope in science fiction.

Does the answer even matter? After all, everything feels quite real. If it's all a simulation, it's a very good simulation. If it isn't real, we can still pretend it is and get on with life.

Does it even matter whether we really had a "real" life in the distant past? If we did, then our present situation is an ancestor simulation and we live in a genuine, religion-free after-life[11] (unless you want to call the sysop or the programmer god, and then consider prayers as attempts at hacking). If we didn't, then we have been constructed *de novo* within the simulation and are therefore AIs. This also answers the question of whether AIs can truly be conscious or creative or alive: we are all three—we insist on it!—and if we are AIs, then...

Now let's get confusing...

If AIs learn to be us by watching, over time they will get better and better at being us. They may even become indistinguishable. So—when our distant descendants get around to simulating us or the AI versions of us, which one will they choose? (Bear in mind that those descendants might be within our simulation, so they will be simulating a simulation. It's turtles all the way down.) It would surely be easier to use just our AIs—they will already be in computer code form, so they'll just have to be tweaked to fit into the simulation. For the nonhuman world, augmented reality will help a lot; Kevin Kelly said at *Wired* on February 12, 2019, that "Someday soon, every place and thing in the real world—every street, lamppost, building, and room—will have its full-size digital twin in the mirrorworld." This too will help ease the creation of future ancestor simulations. It would presumably be much more work to simulate the real us and real world from scratch.

If we live in a base-line reality instead of a simulation, it makes a certain amount of sense to use both us and our nascent AIs, which may (or may not) learn to be us in the next few years. If we live in a simulation, then our descendants will have chosen to simulate both us and our AIs. Unless they used our later AI copies to be us and left the actual AIs for our simulated selves to create.

In a few more years, "our later AI copies" may actually exist. But presumably even later copies will exist in a few more years. And then... Version control is gonna be a bitch!

We need a naming convention: AI for an original one, AAI for a copy of an AI, AAAI for... Wait—that sounds like a repair service for broken-down intelligences.

Enough silliness. The basic question remains: Are we AIs?

I don't feel like one, but then I wouldn't, would I?

So... Damned if I know.

Artificial Intelligence

OH, POO!

No, I have not misspelled an homage to A. A. Milne. Poo, or poop, is a substance of considerable interest to anyone concerned with agriculture (manure is fertilizer) or human health (our intestinal microbiome seems more important every year).

Dirt Needs Dirt

When I was very young, about 1950, my grandparents' house had an outhouse, even though the house was in town. The house, you see, was owned by my great-grandmother, in her nineties, and she insisted that her late husband (d. 1927) would never have approved of such new-fangled things as indoor plumbing, central heating, and electric lighting.[1]

Yet the amenities did reflect a certain practicality. Once a year, a man would come to clean out the privy and haul the accumulated wastes off to a local farm, where they would be plowed under to enrich the soil. Houses with indoor plumbing routed the wastes to a local river. Today wastes flow from the household to a sewage treatment plant to be digested with the aid of assorted microorganisms and turned into "sludge." The sludge is then often trucked to a landfill for burial. It can also be incinerated.[2] Or fermented to produce methane that can be burned to generate electricity.[3]

It can also be applied to farm fields, but urbanites who have moved to the countryside to be closer to Nature often find that agriculture doesn't always smell very sweet. Cows, y'know. And chickens and pigs. And if you even mention the word "sludge," those ex-urbanites complain.[4] Modernity has spoiled them. If they only knew how cities used to smell!

If the truth be told, sludge can indeed stink, though composting can reduce this problem, as well as potential contamination by disease-causing bacteria and viruses. So can plowing it promptly into the soil. You can even eat the food grown in that soil, though you might reasonably prefer crops grown above ground, such as tomatoes, rather than root crops such as potatoes, beets, and carrots.

Animal manure—from cows, chickens, horses, pigs—also stinks, but it is more broadly accepted as fertilizer. The hot-button word that upsets people seems to be *human* manure, as if it were somehow less natural than animal manure.[5] Yet, once upon a time, using human manure was as conventional as using any other kind of manure. It was good for the soil. Dirt needs dirt.

It really doesn't make much sense to bury perfectly good fertilizer in a landfill or to turn it into power plant fuel. It is much better to put it back in the soil it came from in the first place, in the form of plant nutrients such as nitrogen and phosphate.[6] Yet organic farmers have long resisted using sludge on their farms. A major reason is chemical contamination, especially by various toxic industrial chemicals. Industries are supposed to clean up their effluent before they route it into a municipal sewer system, and thence into a sewage treatment plant, but they don't always. If contaminated sludge goes to a farm, chemicals such as lead and mercury can wind up in food or wash off the land into local streams and lakes. It thus behooves any sewage treatment plant to test its sludge for chemical contamination before deciding what to do with it. Landfill or burn the bad stuff. Put the rest to fertile use. Even so, the Environmental Working Group is against using sludge in any food production system.[7]

The liquid that enters a sewage treatment plant is also of interest, for it contains a great many things that have passed through the human body or have been discarded via the flusher. This includes various prescription drugs such as birth control drugs and antibiotics, both of which have been found in the cleaned-up liquids the plant releases to local bodies of water. Birth control drugs in rivers have been shown to affect fish.[8] Antibiotics can lead to antibiotic-resistant bacteria.[9] And since downstream communities may take drinking water from these bodies of water, there is a chance for the drugs to affect humans. Fortunately, they are very much diluted by then, so you don't have to worry about sharing anti-psychotics and other meds. Yes, they're there too.[10]

Sewage effluent also includes illicit drugs such as cocaine and meth.[11] One study found that if you sampled the sewage flow upstream from the sewage treatment plant, you could track drugs, pipe-branch by pipe-branch, to neighborhoods within a city. The researchers did not try to track the drugs all the way to individual households out of concern for privacy.[12] But not everyone has the same attitude. China, for one, appears ready to include waste-stream monitoring for drugs in its extensive surveillance measures.[13]

Given the modern ability to sequence DNA quickly and cheaply, it is surely worth reflecting that where our poop goes, so goes our DNA as cells slough from the lining of the intestine. So far the technique is being used mostly to detect microorganisms, with an eye on monitoring public health,[14] but the potential is either tantalizing or frightening, depending on how you feel about Big Brother watching you.

Imagine, for example, that you know a dastardly criminal is hiding in a city, but you don't know where. You have a DNA sample, perhaps as a bit of skin and blood from under the fingernails of a rape victim. So you check the local sewage treatment plant to see if any of the perp's DNA has made it there. If it has, then you move upstream, sampling the sewage, pipe-branch by pipe-branch, until you are under his toilet. (If the pipes are too small for *you*, send a pipe-crawling robot,[15] perhaps equipped with its own DNA reader.) Then you check your GPS for the address and call the SWAT team.[16]

What could go wrong, after all? Yet a savvy criminal, like a wild animal, might prefer to poop as far from home as possible. Perhaps at a bar or restaurant six blocks from the hide-out. That SWAT team could cause a great deal of excitement. Innocent bystanders could even get killed.

Can sewage effluent be cleaned up so it contains no drugs, DNA, viruses, toxic chemicals, or even the fragrance of its origin? This is a question of some concern to many areas already suffering from droughts, and to even more places expecting future drought due to climate change. Few people have an issue with using sewage effluent to irrigate lawns, golf courses, and farms. But what about putting it back in the water supply system? Well, El Paso, Texas, will be drinking recycled toilet water soon.[17] It's already happening in California, though some of the reuse is "indirect" as cleaned-up effluent is injected into underground aquifers from which wells and pumps can return it to municipal water systems (and homes, of course).[18] Direct reuse is also being considered in Cape Town, South Africa,[19] and indirect reuse in Australia.[20]

Removing viruses and bacteria is relatively easy. Removing toxic chemicals is less easy, but it can be done. Removing drugs is harder, but failure to remove illicit drugs means that repeated reuse can, as people add further rounds of drugs to the water, increase the amount of the drugs in drinking water, perhaps to the point where everyone in town fails drug tests. Or perhaps to the point where prescription drugs start affecting people who don't need them.

The "fragrance of its origin"? That's the yuck factor, right there. No one wants to drink toilet water! Leave that to pet dogs and cats (unless you leave the lid down).

But we are facing a future in which serious droughts are likely to be both more common and more widespread. If the choice is between drinking cleaned-up toilet water and not drinking at all, I know what I'll choose.

The Frenemy Within

An ecosystem is a combination of living species and the environment in which they live. An ecological community is just the living species. There are a great many communities. There are even communities on and within human beings, consisting of the bacteria and fungi—and even the generally harmless follicle mites[1]—on our skin and the bacteria, fungi, viruses, and various parasites in our guts.

Years ago, when teaching an environmental science course, the topic was communities. I assigned the class to write a paper describing one. One pair of students chose to write on the community inside our guts, but handed in virtually the same paper. A little checking revealed that both were copied verbatim from a website peddling herbal remedies guaranteed to keep you from ever seeing in the toilet bowl a long list of parasites. Interestingly, the listed parasites were not ones familiar to a parasitologist. That is, if you did *not* use the herbal remedy, you would never see them either. The particular herbal product doesn't seem to be available anymore, but there are a great many herbal "colon cleansers" for sale. Since, thanks to modern hygiene, parasite infestations are now rare, you don't need to waste your money.

The students of course flunked the assignment—for two reasons. First, they cheated. Second, they were stupid about it—they could have cross-checked what they found. Third—no, I didn't flunk them for this, but third anyway—if they had done a bit of research, they would have found an immense amount of very interesting material on intestinal parasites.

The connection with wearing shoes alone would have been worth a paper.

Say what? It's true—until the early twentieth century, kids in the rural American south commonly went barefoot and infestations of hookworms, pinworms, and other worms were common.[2] Public health workers pushed for wearing shoes (and using toilets or outhouses), and infestations were reduced. So, of course, were the symptoms, which—because the worms consume the food their victims eat—included a decided lack of energy. That lack of energy *in the past* may have more than a little to do with the stereotype of southerners (of all colors) as "slow,"[3] even though worm infestations are now much less common.

Yet internal parasites aren't all bad. Consider the "hygiene hypothesis."[4] It says that our intestinal parasites, fungi, and bacteria have evolved not only to benefit from us but also to provide benefits to us (after all, they depend on us). The microbes comprise our microbiome or microbiota and interact with the immune and other bodily systems in many ways. Parasites appear to as well, for as populations eliminate parasite burdens they become more subject to allergies, asthma, and autoimmune diseases (even, perhaps, multiple sclerosis). One Japanese researcher reportedly infected himself and his graduate students periodically with tapeworms to study their benefits, and other researchers have been trying to use other worms or their components to fight allergies.[5] You can even find do-it-yourself instructions on the Internet (not recommended!).

Most people find the idea of deliberately infecting themselves with—or even tolerating—parasitic worms just a teeny bit repulsive. But there are drugs that can kill worms. Just think of what you do when you find roundworms or tapeworm segments behind your pet dog or cat—

you rush the animal to the vet, where a single pill does the job. In other words, a health-promoting worm treatment could be a short-term sort of thing.

Another implication of the hygiene hypothesis is that we should be less obsessed with cleanliness. Kids that grow up on farms or with pets appear to be less likely to suffer from allergies and asthma.[6] And considering the way pets lick their butts and then lick us, you know they are exposing us to various microbes, or even parasites. Fortunately, these things can actually be good for us.

Little kids also pick up microbes by inhaling floor dust and putting toys in their mouths. Let them. They need the exposure. As my grandparents' generation used to say, we need to eat a peck of dirt before we die.[7]

Buying dish-washing detergent and hand soap with built-in antibiotics is not a good idea.[8] Some of these antibiotics the FDA has actually banned from soap, but there are others. And they don't make the soap any more effective, while they may promote antibiotic resistance, both in the kitchen and downstream in sewage treatment plants and waterways. Since resistant bacteria are harder to kill with antibiotics, if you get infected with them, you can be in trouble.

And it has been suggested that wooden cutting boards, wiped down after use, are more sanitary than plastic cutting boards put through a dishwasher.[9] One hypothesis is that on wood, bacteria are drawn into pores and killed by wood chemicals (such as lignin).

But, but... Yes, we can get sick when we are exposed to the wrong parasites or bacteria. The tropics are rife with schistosomes[10] and guinea worms,[11] among others. There is even a spiny South American fish that follows currents containing ammonia[12]—so don't pee in the river. Dogs have been reported to infect with flesh-eating bacteria by licking.[13] Water parks can give you brain-eating amoebas.[14] Food preparers can contaminate food; the textbook case is "Typhoid Mary."[15] Thousands of tons of food are recalled every year.[16] Medical devices—even implants—can be contaminated, and the consequences can be dire.

The hygiene hypothesis does *not* say that hygiene is irrelevant. It isn't. But it does suggest that we can overdo it. The really scary stuff is rare. Bear in mind that the immune system reacts best to germs it has seen before. It benefits from exposure. So go ahead and retrieve the potato chip that fell on the floor—but pass on the fried egg. The five-second rule is a Real Thing—research has shown that the less time a piece of food spends on the floor the fewer bacteria it picks up, and that moist food picks up more.

Everything in a house does *not* have to be squeaky clean. Maybe not even the people, despite the old saying that "cleanliness is next to godliness." When I was a kid that meant our one bath a week was on Saturday night, in time for Sunday morning church. The rest of the week wasn't that close to godliness, and not bathing daily (except when we'd been rolling in mud) didn't seem to hurt us. It surely still wouldn't, but we have been listening to the soap and deodorant makers for a long time now.

So much of what we take as normal has been influenced by someone's quest for a buck. However, if we only bought what we really needed, the economy would crash.

Your Other Half

In the last few years, the human microbiome has been drawing a great deal of attention. It consists of the bacteria, viruses, and fungi that occupy your guts (generating about a third of your poop), as well as your skin, mouth, nose, and other body zones. Its many species make up a population of microbial cells more numerous than (if not as massive as) your own cells, thus justifying the title of this essay, and they communicate with your brain, immune system, heart, and other organs in ways that have a great deal to do with your health. We're still working out the details, but it is already clear that messing with the microbiome can be very bad for you.

I became aware of the importance of this topic in January 2013, when the *New England Journal of Medicine* published a very interesting paper on the treatment of *Clostridium difficile* infections. These "C. diff" infections often follow antibiotic treatments that destroy the normal intestinal microbe population and cause intractable diarrhea, with resulting dehydration. They affect some 700,000 Americans every year and kill about 14,000.

The researchers planned to enlist 120 patients in the study. Half would get the experimental treatment, half would be split into two control groups. But after 43 patients, the study was halted on the grounds that it would be unethical to continue. Ninety four percent of the experimental group were cured, compared to about a quarter of each control group. "Enough," they said. "Let's give everyone the good stuff!"

If a big pharmaceutical company had a new drug with a 94 percent cure rate, they would be beating the marketing drums and watching stock values soar. In this case, well… Just what was the "good stuff"?

In a word, poop.

Fecal transplantation therapy, now also known as fecal microbiota transplantation or restoration, has been around for a while, but it has been one of those things only the desperate will try. It involves flushing the intestines, more or less the way one prepares for a colonoscopy, and then restoring the intestinal microbes with a preparation obtained by blenderizing a stool sample from a healthy person. The preparation can be administered by enema, or through a tube pushed down the back of the throat to the upper intestine. The technique has been used for C. diff infections, Crohn's disease, irritable bowel syndrome, ulcerative colitis, and other illnesses. The best results show when people who have been ill for months, and whom doctors have been unable to help, leave the hospital feeling great within days. Sometimes a patient may need a second dose from a different donor.

It sounds simple, and it is. Indeed, you can find instructions on the Internet on how to do it yourself.[1] It is of course better to work with medical professionals who can make sure the donor is healthy and the sample holds no hidden surprises.[2] We learned the importance of testing natural products such as blood in the 1980s, when AIDS had appeared but no one yet knew it was caused by a virus, HIV. Writer Isaac Asimov was only one of many people who developed AIDS from blood transfusions. Today, donated blood is routinely tested for HIV and many other viruses and parasites. A similar approach to stool samples would be wise, and some doctors are already stockpiling frozen stool samples that have tested clean. The need for such practices was recently underlined when the FDA issued an alert because antibiotic-resistant bacteria in fecal transplants had caused infections and one death. The FDA is now requiring that fecal transplants be screened for antibiotic-resistant bacteria.[3]

In the C. diff study, the experimental treatment was precisely a fecal transplant. In one control group, patients received the potent antibiotic vancomycin. In the other control group, patients had their intestines flushed out (but not "rebooted") and were given vancomycin. So for C. diff, poop is better than antibiotics.

But it's still poop, and the ICK! or YUCK! factor is very strong in most folks' responses to the idea of using it. At least until they come down with one of the conditions for which an intestinal reboot has been shown to work and grow desperate enough to try anything.

Is it possible to take the ICK! out of it? Several groups are working on isolating the important bacteria and culturing them to make a poop-less preparation that amounts to a super-probiotic. In Canada, one project is called RePOOPulate.[4] In Minnesota, a company called Rebiotix is developing a "microbiota restoration therapy" for C. diff cases; the Food and Drug Administration (FDA) fast-tracked their phase II clinical trials, which produced very positive results; phase III trials are under way now. In Cambridge, Massachusetts, Seres Therapeutics has developed an algorithmic approach to identifying and concentrating poop bacteria that may correct conditions such as C. diff infections. Their preparation was given "breakthrough" status by the FDA, and to raise money for Phase III trials, they went public in April 2015. Investors hoped to do well.

How well? Consider a paper by Vanessa K. Ridaura, *et al.*, "Gut Microbiota from Twins Discordant for Obesity Modulate Metabolism in

Mice," *Science,* September 6, 2013. They chose eight women, four twin pairs who were "discordant" for weight (in each pair, one was heavy, one was not), and transferred fecal bacteria from each woman into germ-free mice. Mice that received bacteria from the obese twins gained weight. Mice that received bacteria from the lean twins did not. And to top it, when the resulting fat and thin mice were caged together, the "lean" bacteria infected the fat mice (mice eat each other's poops), which then lost weight.

Would some version of this procedure help humans lose weight? I'm sure that entrepreneurs are drooling at the thought that it might (weight-loss products are a gold mine, especially if they work), but there is nothing in this study that says it would. Mice are different from people. Still, there have been a number of studies that indicate that obese and lean people have different gut bacteria populations and manipulating those populations may do some good.[5]

The entrepreneurs and investors thus may well have something to drool over. But a good deal of research remains to be done before we see fecal transplants (or bacterial preparations derived from stool samples) available to treat obesity as well as debilitating intestinal conditions. That research is already under way, with researchers testing fecal transplant therapy on many diseases.

Among those many diseases are some that outfits like Rebiotix and Seres are already working on. They include Crohn's disease, ulcerative colitis, irritable bowel syndrome, metabolic syndrome, multidrug resistant urinary tract infections, and hepatic encephalopathy. Seres is also developing preparations to help cancer immunotherapy be more effective and prevent infections and immunity issues after organ transplants.

A little further in the future, there may be hope for preventing at least some cases of autism. In 2018, Catherine R. Lammert, et al., of the University of Virginia published a report indicating that a major autism risk factor may be the condition of the mother's intestinal microbiota.[6] There is also evidence of a possible link with depression.[7] Because the microbiota can be modified easily by changing diet or administering probiotics or even fecal transplants, it may prove possible to reduce the risk of both autism and depression.

In 2017, W. H. Wilson Tank and Stanley L. Hazen published an assessment[8] of the possible link between the intestinal microbiota and cardiovascular conditions such as atherosclerosis, noting that though a great deal of work remains to be done, "novel diagnostic, therapeutic, and preventive strategies that leverage their identification may become part of our arsenal for halting and reversing cardiovascular diseases."

The day may thus not be far off when the drug of choice for many diseases is a preparation derived from poop and designed to reset the intestinal bacterial population to a pattern more consistent with good health, or even with the good health of the baby in the womb

It will surely help if we can either learn to hold our noses or get rid of the ICK! factor. Using cultured bacteria and wrapping them up in "super probiotic" capsules is surely the better way to go.

But it still comes from your other half.[9]

Oh, Poo

EXTREME TECH

Anyone interested in the future has to be interested in what kind of technology is under development today. This can be genetic engineering, self-driving cars, improved nuclear power, and a great deal more (some of which was covered in the AI section).

Making It Up

What comes after genetic engineering?

Ever since its origin in the 1970s as "recombinant DNA," some people have wrung their hands and cried alarm about how swapping and editing genes (made of DNA or deoxyribonucleic acid) in our food posed hazards of toxicity and ecological destruction. If we ever reached the point of doing the same with people, they cried, we would no longer be human!

Of course, the technology moved forward, and it has in fact been applied to humans. It hasn't always worked, and there have been some very unfortunate outcomes (meaning deaths). But the work has continued.

So far the intent has been to correct genetic disorders (inherited diseases), with care being taken to keep modifications out of tissues that can produce sperm and eggs and thus send the changes on to future generations. Patients, by and large, seem to be in favor of fixing their diseases. Post-humanists are also in favor, for they imagine numerous ways the human body can be enhanced, from genetic redesigns to cyborgish implants (see the Humanity+ website[1]). Critics of the technology are distinctly less enthusiastic; they remain concerned that the technology will be used to change the basic human template. But Harvard Medical School's George Church has quipped "What is the scenario that [they're] actually worried about? That it won't work well enough? Or that it will work too well?"[2]

And then something hit the fan. Just before the 2018 International Summit on Human Genome Editing, held in Hong Kong, China, He Jiankui of the Southern University of Science and Technology in Shenzhen, China, made headlines by announcing that he had genetically engineered twin girls who lack (at least in part) a gene that permits HIV, the AIDS virus, to infect human cells. The kids are just babies now. If they grow up without ill effects from the gene tweak and have kids of their own, they will pass the change on. Dr. He's goal, to figure out and test a defense against a nasty virus, is laudable. If the defense works and is widely implemented, a redesigned (in a small way) humanity can sigh with relief. But he was greeted with widespread dismay and condemnation, as well as by calls for better international regulation, perhaps through the United Nations.[3] By January 2020, he had been sentenced to three years in prison.[4]

It's a kerfuffle. For Dr. He, it's worse than that. Other researchers may also be reflecting on what such work could do to their career prospects. But though such responses may slow progress down a bit, they won't stop it.

So what's next?

They call it "synthetic biology." Instead of involving the moving and editing of genes, it involves making up new ones and putting them into bacteria, plants, animals, and maybe even, someday, people. And yes, it scares some people spitless.[5]

It began two decades ago, when—just to prove they could do it!--researchers constructed a live poliovirus from simple lab chemicals. A decade later, researchers assembled a bacterial chromosome, stuck it in a bacterial cell that had been stripped of its genes, and Lo! It worked. Ultimately, researchers want to go much further, even to the point of creating synthetic organisms out of chemicals and totally artificial DNA, rather in the way kids build things out of Legos. There is even talk of making a new kind of genetic material (XNA—the X stands for "Xeno") that doesn't use the same components (nucleotides) as DNA.[6]

The goal is to exert unprecedented control over what cells do. In testimony before the House Committee on Energy and Commerce Hearing on Developments in Synthetic Genomics and Implications for Health and Energy, May 27, 2010, Craig Venter said "The ability to routinely write the 'software of life' will usher in a new era in science, and with it, new products and applications such as advanced biofuels, clean water technology, food products, and new vaccines and medicines. The field is already having an impact in some of these areas and will continue to do so as long as this powerful new area of science is used wisely."

Allen A. Cheng and Timothy K. Lu, "Synthetic Biology: An Emerging Engineering Discipline," *Annual Review of Biomedical Engineering*, August 2012, see synthetic biology as bringing the engineering mindset to biology, with great potential for human health. Timothy K. Lu and Oliver Purcell, "Machine Life," *Scientific American*, April 2016, discuss building into cells the biochemical equivalents of computer circuits so the cells can detect and treat illness and environmental pollution. There is even industrial potential. The Defense Advanced Research Projects Agency (DARPA) has launched a program called Living Foundries "to create a new manufacturing capability for the United States" using synthetic cells.

Not surprisingly, some people think this kind of work challenges the divine monopoly on creation. Others fear that if one can construct one virus from scratch, one might construct others, such as the smallpox virus, or even tailor entirely new viruses with which natural immune systems and medical facilities could not cope. New bacterial diseases become possible as well. And there is the potential for new kinds of terrorism. In

2010, Vatican representatives declared that synthetic biology was "a potential time bomb, a dangerous double-edged sword for which it is impossible to imagine the consequences" and "Pretending to be God and parroting his power of creation is an enormous risk that can plunge men into barbarity."[7] As I was writing this essay, Vint Cerf published "Synthetic Organisms Are about to Challenge What 'Alive' Really Means."[8] The definition of life isn't as fragile as he seems to think, but his conclusion—"In 2019 we will need to begin a serious debate about whether artificially evolved humans are our future, and if we should put an end to these experiments before it is too late"—is very much on the "be cautious" side.

The "playing God" objection seems likely to grow louder as synthetic biology matures, but it is also likely to fade just as it has done after previous advances such as in *vitro* fertilization and surrogate mothering. According to one study, the general public already tends to approve of synthetic biology work aimed at "societal, medical, and sustainability needs."[9] There is also a great deal of interest from investors.

Is the fuss premature? Not much has actually been accomplished so far. One gets the feeling that researchers are doing this stuff just because they think they can. But a team of plant scientists has inserted novel genes into tobacco plants to make photosynthesis more efficient. The plants grew 40 percent bigger, and the technique looks like it could be applied to food crops to increase production.[10] And while this was "just" genetic engineering, not synthetic biology, it does speak to the potential value of being able to modify living things.

Should we be worrying about the potential impact of synthetic biology? It might, of course, be misused. Eventually, it will surely be applied to human beings, first to cure or prevent disease, then to improve function, and *then* to add whole new features to the human package (Transhumanists, rejoice!).

But isn't it really just genetic engineering? Wasn't that "engineering mindset" already there before anyone coined this new label? Sure, but maybe they thought calling it "extreme genetic engineering" sounded more like something for a sports channel.

Hmm. Maybe someday.

It's Loose!

What's loose?

If you're a fan of old movies, you are probably thinking Giant BUGS! Ants! Moths! Lizards! Squids! Things from the swamp! And of course those velociraptors from *Jurassic Park*.

You're partly right. Bugs, but not giants. Genetically engineered (or gengineered) bugs, with features Mother Nature never dreamed of, unless she was having a bad night.

To be more precise, mosquitoes gengineered to make their offspring either sterile or dead. (Ma Nature wouldn't dream of that because Her whole point is reproduction.)

Why would the gengineers want to do that? Well, for one thing, mosquitoes are the deadliest animals on the planet (other than us). They spread diseases—malaria, yellow fever, dengue, Zika, West Nile, chikungunya. And the gengineers have said to themselves, "We can stop that."

They're not trying to do something really, really cool like bringing *Tyrannosaurus rex* back from extinction. (But hold that thought.) They're trying to be useful, and in a big way.

Aren't there other ways to solve the problem? Vaccines, maybe? We only have a good one for yellow fever so far. There's one for dengue, but it can actually make the disease worse. Insecticides? Unfortunately, insecticides create more problems than they solve. They kill a great many insects—predators, bees, butterflies—besides the ones you're after. They are also toxic to humans and other animals. And natural selection ensures that insect populations become resistant to insecticides, so the chemical approach never works for long.

The sterile male technique was devised in the 1950s to control the screwworm fly, which lays its eggs in open sores on the hides of cattle. The hungry larvae can kill cattle within 10 days. The technique involves raising large numbers of bugs in the lab, exposing them to radiation to sterilize the males (and the females, but the males are the point), and releasing them wherever there is a screwworm problem. Many or most wild females mate with a sterile male, and—since they mate just once and store the sperm—their eggs are then duds. The population crashes. This is how the screwworm was eliminated as a problem in North America and Africa. The technique has also been applied to other insects such as the Mediterranean fruit fly and mosquito, though with less dramatic success because the females mate more than once.

The World Mosquito Program,[1] is working intensively on using a bacterial symbiote, *Wohlbachia,* which research shows can reduce the ability of mosquitoes to pass on dengue, chikungunya, malaria, and yellow fever. They propose to release *Wohlbachia*-infected mosquitoes in disease-prone areas and have already done so in Australia and Miami; results are encouraging.

There are several approaches to using gengineering to attack the problem. One modifies specific genes to make male mosquitoes sterile (instead of exposing them to radiation). A second approach introduces new genes to make offspring die before they can mature and transmit disease. A third approach fiddles genes to make almost all offspring male. All three aim to reduce reproduction of mosquitoes in an area so that their numbers and disease transmission both decline. A fourth approach aims to make mosquitoes less able to carry and spread disease.

Of course there is opposition. The anti-genetic engineering folks think it's unnatural, it infringes the genetic integrity of species, and who knows what awful things will happen if the engineered genes get into other critters, like maybe even people! Adrienne LaFrance, "Genetically Modified Mosquitoes: What Could Possibly Go Wrong?" *Atlantic,* April 26, 2016, recognizes the need to control mosquito populations and notes that many objections to genetic engineering are rooted in misinformation. In other words, no, genes that sterilize mosquitoes won't sterilize people, or hummingbirds (which do eat mosquitoes). Fear of disease, she says, is likely to mean genetically engineered mosquitoes will eventually be widely released into the environment.

So far, most genetically engineered mosquitoes have been tested only in laboratory cages. However, the British company Oxitec has released genetically engineered male mosquitoes whose offspring die. They have reported success in the Cayman Islands in the Caribbean (2009) and in Brazil (2013). Activists are trying hard to prevent approval, but so far without success.

It's only a matter of time before these mosquitoes are released wherever mosquito-borne diseases are a problem. Eventually, we might even release them where the only problem is itchy mosquito bites.

And if mosquitoes go extinct? Their larvae are fish food, and they themselves are bird food. But nothing (as far as we know) eats nothing but mosquitoes and is in danger of extinction if we do away with the bugs.

Coming soon... Maybe...

Gengineers are also working on pigs that don't carry the remnants of ancient retroviruses. If they succeed, people who need organ transplants could get them from pigs instead of motorcyclists who don't wear helmets. They're also working on cattle that produce beef more efficiently and lambs that are meatier. If they succeed, the big question will become whether people will accept the fruits of their work. Some, at least, are sure to become willing steak-holders.

A little further down the road is gengineering to bring the mammoths back. The engineers know the recipe, for they have DNA samples. What they need to do is edit elephant DNA to match mammoth DNA and let elephants bear and raise baby mammoths. If successful, they can then release mammoths into the North American wild, where their grazing may help the land store more carbon and thus fight climate change.[2]

If you're interested in recreating an ancient ecosystem, you also need predators such as dire wolves,[3] though they apparently preyed on horses more than mammoths. We may not need gengineering to bring them back, however. Dog breeder Lois Schwarz is trying to "breed back" from dogs and wolves to get something that at least looks like a dire wolf; she is the founder of the Dire Wolf Project.[4] If she succeeds, the gengineers might need to make just a few tweaks to add in an appropriate level of ferocity.

Nothing could possibly go wrong, right?

It would surely be safer to bring back the passenger pigeon, and in fact there is a Great Passenger Pigeon Comeback Project[5] that will require a good deal of gengineering.

But if you want to do birds—and the passenger pigeon really is an excellent choice for we have tissue samples and it's a conservation icon—how about the terror birds?[6] The Florida version, *Titanis walleri*, reached North America about 5 million years ago and lasted for nearly four million years. It stood about 7 feet tall, weighed over 300 pounds, had a beak the size of a suitcase, and was a thoroughly impressive predator.[7] DNA has been recovered from fossil eggshells, but Google finds no trace of revival projects.

Perhaps that is just as well. It probably wouldn't have much interest in your birdfeeder. Even your German shepherd would be but a snack. It would find more satisfaction at the Costco meat counter. Or maybe the checkout line or parking lot.

Still... We do have some DNA to play with. If we can do mammoths and passenger pigeons and tweak dire wolves, we can probably do terror birds in a decade or so. However, I can just imagine the cries of horror from the anti-genetic engineering folks. Not just unnatural! And genetic integrity! But also WE'RE ALL GONNA DIE!!

Sounds like *Jurassic Park*, doesn't it? That's not possible since we don't have dinosaur DNA samples. But a park with mammoths and dire wolves and terror birds *would* be possible. We could even turn them loose on the landscape.

But... WE'RE ALL GONNA DIE!!

Human beings are the most dangerous creatures on the planet. We could probably change the minds of at least some of the anti-genetic engineering folks just by saying "hunting."

Just imagine a game park with terror birds!

"Hold my beer, will you? I've got that!"

And now that I've offended the anti-hunting folks, I'll stop.

Making Us Better

Not long after the first baby steps toward genetic engineering (or gengineering) were taken in the early 1970s, critics such as Andrew Kimbrell tried to jump down the field's throat. In his eventual book, *The Human Body Shop: The Engineering and Marketing of Life* (HarperSanFrancisco, 1993), he said the development of genetic engineering was so marked by scandal, ambition, and moral blindness that society should be deeply suspicious of its purported benefits.

Of course, that did not stop progress. Almost instantly, the gengineers were making bacteria and yeast produce drugs such as insulin and growth hormone with zero risk of contamination by prions such as those that cause Creutzfeldt-Jakob and Mad Cow disease. Then came gengineered crops with added bacterial or other genes to make them toxic to insects (but not to us) or resistant to herbicides and various diseases.[1] Most recently, a Chinese researcher has stirred controversy by engineering two babies to lack a gene that helps the AIDS virus (HIV) infect cells; the modification may also have made the babes smarter.[2]

Now we're getting into *Body Shop* territory. If you could gengineer for smarts, how many parents would sign up? What else would they sign up for? Lack of cancer susceptibility genes such as BRCA?[3] Lack of other disease susceptibility genes (heart, autism,...)? Stronger immune system? Immunity to tuberculosis or flu? Physical attractiveness? Height? Strength? Better vision or hearing? Creativity or talent in some area? All the above? The list is long, though we have no clue as to the genes we might engineer for some traits, even when we know that, for instance, a talent for pastry making or singing or acting might run in the family.

How about redesigning the body? A prehensile tail might be handy for holding car keys while trying to open a door. It would also let one get up to monkey business under the table. The ability to regrow lost body parts would be a boon to a soldier or a motorcyclist. Photosynthetic skin would save money on lunch breaks. Many folks would love spinal disks that don't rupture, joints that last longer, or even just the ability to burn more calories faster.[4]

And it's not just genes. Researchers are already testing brain implants to restore the ability to move to quadriplegics[5] and even to turn brain waves into speech.[6] On the input side, one implant under development promises to help the blind see.[7] There are also auditory brainstem implants.[8] Add memory boosters,[9] and you're awfully close to being able to construct a whole internal PC. And then there are implantable chips to help restore brain functions lost to concussions or Alzheimer's Disease.[10]

The very idea of changing the human body, whether with gengineering or with brain implants, horrifies many people. They fear that such changes will make us no longer human. Stephen Greenblatt, a Harvard humanities professor, argues that our biological flaws are an essential part of what makes us human.[11] He says, "there is a huge gap between, say, repairing spinal cord injuries by implanting electrodes in the brain and implanting an intelligence-enhancing AI chip... The one restores an injured human to full mobility; the other alters the very nature of the human, at least as it has been conceived for millennia across a substantial part of the globe."

Is "the way it was" somehow holy? Greenblatt seems to think so, but others have a very different attitude. There is in fact a movement known as transhumanism that "promotes an interdisciplinary approach to understanding and evaluating the opportunities for enhancing the human condition and the human organism opened up by the advancement of technology" (see http://humanityplus.org/). The tools are gengineering and computer implants, and the goal is to eliminate aging, disease, and suffering. The transhumanist vision extends to "post-humanism," when what human beings become will make present-day humans look like chimpanzees by comparison.[12]

Transhumanists scare the socks off the traditionalists. Francis Fukuyama,[13] has called transhumanism "the world's most dangerous idea." He has a point, for if the various improvements are available only to the rich, there is a likelihood of creating a new upper class, *a la* the movie *GATTACA*. For this reason, Maxwell J. Mehlman[14] has argued that the day of routine enhancements is coming and there is no real hope of banning the technology. Therefore, it must be regulated and even subsidized to ensure that it does not create an unfair society.

Whether we ensure that everyone has access to this technology or not, there will be a need to learn how to live with it. Prehensile tails, for instance, will require a whole new chapter in etiquette books and websites.[15] A brain chip with Wi-Fi or Bluetooth opens up a whole new category of privacy concerns: If a chip can translate brain waves to speech or text, a hacker from the CIA, NSA, FBI, local law enforcement, a foreign intelligence agency, or a merchant could eavesdrop on your thoughts.[16] If it can go the other way—and it will have to, since one-way Wi-Fi isn't much use—they may be able to control—or at least influence—your thoughts (ads *behind* your eyeballs!). You can bet that high-security jobs will ban these chips. Politicians won't want them either, since the news media would love nothing better than access to a President's very thoughts. It would be even better than his tweets!

A Wi-Fi-connected computer in your skull would change other things as well. School exams would no longer be able to test memory (Google is a wonderful extra memory bank, and you can look up everything ahead of time in case the teacher blocks your Wi-Fi—see next paragraph). Reading could become a lost art as books are copied directly to your memory. Mobile phones and tablets would go the way of the buggy-whip. And more.

If the chip is inside your head, you may not be able to turn it off. But electromagnetic waves (radio, or Wi-Fi) can be blocked by Faraday cages, and one UK bar owner built one to keep customers from using their cell phones.[17] Tinfoil hats (get yours here![18]) may become obligatory fashion accessories instead of the favorite method for crackpots to keep aliens out of their heads. Perhaps unfortunately, if politicians wore this sort of haberdashery, that would only confirm the condition of their crockery.

Gengineering and implants are only the beginning. Transhumanists tend to look forward to the day when computers have enough memory to hold a human mind, and of course to the ability to copy a mind into the computer. Ray Kurzweil[19] and other futurists have been predicting we will be able to do this by around mid-century, but it may take longer. On the other hand, there *is* already a startup.[20] If this happens, there is virtually no limit to what your body could look like. If you're in the computer, you can be installed in a robot, a truck, a cruise liner, a spaceship... Why use an

Artificial Intelligence when you can have a Real Intelligence?

Oh... Wait... One of my earlier essays said we might actually *be* Artificial Intelligences. If that's so, it should make uploading minds a lot easier to achieve.

The transhumanist future sounds rather exciting, doesn't it?

It doesn't? Sorry. As Mehlman says, it looks inevitable. If you don't embrace it, your kids will, just as you have embraced massive changes since your parents' and grandparents' time.[21]

How Safe Is Safe?

One of the foundations of computer science and thus of all the gadgets that enrich (or bedevil—take your pick!) our lives is cybernetics. The root is the Greek word for "steersman" (as of a ship) and it is apt for the study and implementation of electrical devices that produce appropriate outputs in response to changes in incoming data. A very simple example is a household thermostat that turns the heat on when the temperature drops and off when the temperature rises. A nonelectrical example is a toilet tank that turns the water on when the level of water in the tank drops (as after a flush) and then turns it off when the level in the tank returns to the full mark.

You can easily think of other examples where changing inputs require changes in outputs, and perhaps where the word "steersman" might be apropos. Driving a car would be one. We have to stay on the road when it curves, slow down when another car is in front, stop at red lights, start up again when the light changes, stop quickly when pedestrians step into the road, avoid idiots in other cars, and so on. It's a complex task, but even teenagers can master it—at least well enough to get a driver's license. Unfortunately, even people with years of experience make mistakes, have accidents, and kill people. Indeed, driving causes some 35,000 deaths per year in the United States alone (2015 data).

It should thus surprise no one that making cars safer has been on the cybernetic and computer science to-do list for a long time. The ultimate goal has been cars that drive themselves, with no drunken, cell-phone-using, newspaper-reading (I've actually seen that!) human klutz at the controls. They're called autonomous cars or robocars.

In 2004, 2005, and 2007, the U.S. Defense Department's Advanced Research Projects Agency (DARPA) held "Grand Challenge" and "Urban Challenge" competitions for early versions of the technology. In 2014, "The DARPA Grand Challenge: Ten Years Later"[1] reported that the events "created a community of innovators, engineers, students, programmers, off-road racers, backyard mechanics, inventors and dreamers who came together to make history by trying to solve a tough technical problem." The result was rapid progress toward driverless delivery bots, taxis, trucks, and cars by the likes of Uber, Waymo, Tesla, GM, Toyota, Volvo, and more. Performance so far has been encouraging, but the driverless cars produced so far *have* had accidents. As Nidhi Kalra of the RAND Corporation noted in his testimony before the U.S. House of Representatives Committee on Energy & Commerce, Subcommittee on Digital Commerce and Consumer Protection, February 14, 2017, the technology will not—indeed, cannot—eliminate all accidents. Bad weather, for instance, will always be a problem. One recent report indicated that the cars may have trouble seeing dark-skinned pedestrians.[2] And self-driving cars, being intensely computer-based systems, will be susceptible to problems all their own, such as (computer) crashes and hacking. Nevertheless, they hold immense potential to make driving safer. After all, they can't get drunk and they don't get distracted by cell-phones.

Not all self-driving cars are equal. The ultimate goal is a car that can do its job in any weather, any lighting conditions, any place, any time. That capability is known as Level 5 autonomy. This is what you want when you stumble out of a bar too plotzed to drive, crawl into the back seat, and say "Home, James." We have a long way to go.

Level 4 autonomy is limited in terms of speed, weather conditions, location. This is the goal for the next few years. Level 3 is similar, with the added need to tell when the attention and skills of a human driver are needed and to sound an alert. Level 2 requires a human driver on deck at all times, ready to take over at a moment's notice. Some near-Level 2 cars are already available, and they have a serious problem in that human drivers think their cars are more autonomous than they really are. This has already led to a few "Oops" moments.

Level 1, we've had for a while. Level 1 cars have at least cruise control. Some can stay in lane, or help with parking. They're not self-driving, but they're helpful.

Testing how well a car of any level works means racking up road miles. Yet it may not be possible to rack up enough test miles to satisfy lawmakers, insurance companies, and customers without actually putting the cars on the market. That might work, for fewer consumers now say they'd be afraid to ride in a Level 5 car (just 63 percent in 2017, versus 78 percent the year before).[3]

Why would they be afraid? Well, duh. Is it safe? How safe? How safe is safe? "When It Comes to Self-Driving Cars, What's Safe Enough?" asks Maria Temming in Science News, November 21, 2017. Human drivers cause 1.1 deaths per 100 million miles of driving. A 10 percent improvement in safety (to one death per 100 million miles) would save 500,000 lives over 30 years.

Is it enough that a self-driving car be as safe a driver as a teenager? One of the DARPA Urban Challenge contestants was reportedly good enough to qualify for a California driver's license. Or would you rather have a self-driving car as safe as the average driver?

Want something even safer? Is Temming's 10 percent improvement over the average driver good enough? Perhaps not, for Temming does note that "people may be less inclined to accept mistakes made by machines than humans, and research has shown that people are" more likely to accept risks they can control. Or *think* they can control. None of us are as good behind the wheel as we think we are. And then there are all the other idiots on the road.

It would be interesting to run a poll that asked people how they would feel about putting those other idiots in self-driving cars. I suspect they would agree that would make the roads much safer. However, I recall one of my father's basic instructions: "Drive like everyone else on the road is an idiot. And don't be too sure about yourself."

The question then becomes how to persuade everyone to adopt the new technology. I suspect that once enough self-driving cars are in use and the insurance industry has accumulated enough statistics, rates will go down for self-driving cars and/or up for manual cars. If the safety statistics are convincing enough, lawmakers may well decide to ban manual cars in cities. Eventually, the ban may be total.

That thought already has some people feeling alarmed. Charles C. W. Cooke, "The War on Driving to Come," *National Review,* December 18, 2017, thinks that the technology will mean the loss—and even banning—of human-driven cars and thereby infringe on his right to drive himself around. People are even saying "They will have to pry my steering wheel from my cold, dead hands," quite as if the Second Amendment were not just about guns. Both truckers and taxi drivers fear the competition will put them out of work.[4] And there is a Human Driving Association, complete with manifesto, at http://humandriving.org/.[5]

Will self-driving cars happen anyway? Of course they will. The only real question is when. And the answer to that very much depends on what we decide about how safe is safe enough.

Extreme Tech

THE CONSTRAINTS WE FACE

How many of us can there be? How rich a standard of living can we all have? What can we do? How far and how fast can we go? How long can we last? We would love to think there are no limits, but there are. The world is finite, and that constrains us.

Are There Limits?

In November 2018, the Trump Administration released Volume 2 of the National Climate Assessment. CNN's headline was "Climate Change Will Shrink US Economy and Kill Thousands, Government Report Warns."[1] President Trump, however, continued to insist that climate change was just a hoax and to undo protections put in place under previous administrations. In March 2019, the United Nations released its sixth Global Environment Outlook[2] to say that thanks to human activities the "ecological foundations of society" and human health are now seriously endangered. In May 2019, an Australian report warned that many of the world's societies face an "existential risk" by 2050.[3]

Almost 50 years ago, in 1972, Dennis Meadows, an MIT social policy analyst, and his colleagues published *The Limits to Growth: A Report for the Club of Rome's Project on the Predicament of Mankind* (Universe Books). The Club of Rome was an independent, international group of scientists, economists, humanists, and industrialists who shared the view that traditional institutions and policies could no longer properly evaluate the complex, major problems that the world was facing. The Meadows group used a computer model (primitive by today's standards) to study five factors that might limit global growth and development: population, agricultural production, natural resources, industrial production, and pollution (now understood to include excess carbon dioxide). The conclusion was that if current policies of unrestrained growth were allowed to continue, without some conscious effort to achieve a state of developmental equilibrium, the next century would likely see a disastrous decline in population and industrial productivity, resulting from resource depletion and widespread pollution.

In other words, the economy would shrink and thousands would die.

The report was greeted by enormous skepticism. Nobody wanted to believe it. After all, the world had been chugging merrily along for a long time with no problems, and never mind that as early as 1864 George Perkins Marsh, in *Man and Nature,* had cried that "We are, even now, breaking up the floor and wainscoting and doors and window frames of our dwelling, for fuel to warm our bodies and seethe our pottage, and the world cannot afford to wait till the slow and sure progress of exact science has taught it a better economy." Nothing had changed by 1948 when William Vogt's *Road to Survival* warned that environmental and overpopulation problems would in time mean hunger, disease, and even civilizational collapse.

The more extreme critics falsely claimed that *Limits to Growth* predicted the collapse of civilization. It didn't, but today some well-known environmentalists are doing precisely that. Sir David Attenborough, speaking to a UN climate meeting, said, "Right now we are facing a man-made disaster of global scale, our greatest threat in thousands of years: Climate change. If we don't take action, the collapse of our civilizations and the extinction of much of the natural world is [sic] on the horizon."[4]

People sound surprised at such warnings today, almost half a century after *Limits to Growth* was published. But they shouldn't. In 1992, the Meadows group published *Beyond the Limits,* a 20-year update of the original study using a much-improved computer model. Their conclusion at that time was that future prospects had not improved and that, in fact, "humanity had already overshot the limits of Earth's support capacity." In 2002, the 30-year update did not alter that conclusion. Nor did the 40-year update in 2012, Jorgen Randers' *2052: A Global Forecast for the Next Forty Years (A Report to the Club of Rome Commemorating the 40th Anniversary of The Limits to Growth)*. Randers argues that humanity's ecological footprint[5] is too large. That is, humanity uses more resources than one planet Earth can supply. Between now and 2052, there will be efforts to reduce the footprint, but they will be too little and too late, in large part because of short-term thinking by governments and businesses. As a result, economic growth and consumption will slow, and the developed world will decline, but resource and climate problems will not become catastrophic until after 2052. He does not trust technology to save the situation. It certainly hasn't shown any signs of doing so yet; in 2019, the world used up the amount of resources (such as fresh water and topsoil) that the world generates in one year by July 29, the earliest "Earth Overshoot Day" ever and two months earlier than in 1997.[6]

In 2008, Graham Turner, a senior research scientist at Australia's CSIRO Sustainable Ecosystems program, published "A Comparison of *The Limits to Growth* with Thirty Years of Reality," Socio-Economics and the Environment in Discussion, CSIRO Working Paper Series 2008-09 (*Global Environmental Change*, August 2008), and found that the projections and the data show a remarkable match. Indeed, he said, lacking immediate action, global collapse before the middle of this century seems far too likely. The picture had not changed by 2014, when Turner published "Is Global Collapse Imminent?" Melbourne Sustainable Society Institute, Research Paper No. 4.

In case you're wondering whether just maybe those old computer models were as bogus as critics complained, in 2019 Zeke Hausfather and colleagues analyzed computer models used to predict the effects of rising carbon dioxide levels all the way back to 1970 and found them spot on.[7] Hausfather, et al., had to do a bit of tweaking, but not to the models themselves. For example, "a famous 1988 model overseen by then–NASA scientist James Hansen... predicted that if climate pollution kept rising at an even pace, average global temperatures today would be approximately 0.3°C warmer than they actually are. That has helped make Hansen's work a popular target for critics of climate science." However, say Hausfather, et al., "most of this overshoot was caused not by a flaw in the model's basic physics [but] because pollution levels changed in ways Hansen didn't predict. For example, the model overestimated the amount of methane—a potent greenhouse gas—that would go into the atmosphere in future years. It also didn't foresee a precipitous drop in planet-warming refrigerants like some Freon compounds after international regulations from the Montreal Protocol became effective in 1989. When Hausfather's team set pollution inputs in Hansen's model to correspond to actual historical levels, its projected temperature increases lined up with observed temperatures."

In a way, this does support one of the criticisms of computer models: "Garbage in, garbage out." Bad data going into the model can produce bad output. But Hausfather's team shows that it's not the model's fault. If you improve the data, the results improve.

We have understood the science for a long time. We have known that this day—when collapse might be imminent—was coming for a century and a half. It's not news anymore. It's long past the time when we should have gritted our teeth and buckled down to solving the problems.

It's not all about global warming or global climate change, even though that's what I started this essay with and it does indeed dominate our current fears. To limit how bad climate change gets, we need to reduce our emissions of greenhouse gases, chiefly carbon dioxide. That means, among other things, stopping our use of fossil fuels. If we can't do that, we may be able to use geoengineering—adding materials such as water droplets or sulfates to the upper levels of the atmosphere to reflect solar energy before it reaches the ground. Unfortunately, we don't know if it will really work or if it will have horrible side-effects. Further, if we start doing that and we do not stop emitting carbon dioxide, we can never stop. See "Throwing Shade," below.

Is there anything else we can do? Of course, since it's not all about climate change. There's population, which is vastly greater than the long-term carrying capacity of the planet. If we could reduce it—or, failing that, if our trashed environment could reduce it for us—our carbon emissions would go down, even if *per capita* emissions did not. So would our other environmental impacts. Deindustrializing would have similar effects. These are messages stressed by the Degrowth Movement[8] and underlined by the 2020 Covid-19 pandemic, which taught us that when everything was shut down, air quality improved everywhere.

Could religion help? It is cursed with conflicting messages. Is the Earth for humans to exploit *ad libitum*? Some actually say the Second Coming and the Rapture are imminent and we should use the Earth up first, since after that it won't matter. Or are we supposed to be stewards of the Earth, preserving it for future generations? If we get it wrong, will God whop us upside the head to correct us? Or will Mother Nature, the Earth Goddess Gaia, or the Earth itself do the job? On April 8, 2020, Pope Francis said that the Covid-19 pandemic might be nature's response to, if not actual revenge for, global climate change.[9]

Many people find the ideas that there are limits to the growth of human civilization, that the way we organize our political and economic systems is fundamentally flawed, that we must reduce population and economic activity, and even—in the name of social justice—redistribute wealth, repugnant. This was already apparent when the original *Limits to Growth* study was published in 1972. Conservative critics insisted that the marketplace would work to prevent disaster with no need to impose international plans or controls. Liberal critics argued that restraints on growth would hurt the poor more than the affluent. And radical critics contended that the results were only applicable to the type of profit-motivated growth that occurs under capitalism. But later studies using more sophisticated models have reinforced the basic conclusion that growth without limits under any economic or political system will ultimately result in disaster. The growth fetish is the problem. We need an ethic of *sufficiency.* Growth should stop when people have enough for their *needs.* Yet we have an entire industry dedicated to convincing people to expand their *wants* endlessly. I mean, of course, the advertising industry.

As a biologist, I do not find the idea that there are limits to growth at all surprising. Indeed, I find economists, politicians, and even preachers who insist that there are no limits so blind, so foolish, that they are guilty of crimes against humanity. We cannot increase population endlessly. We cannot turn every scrap of the planet into product. To insist that we can is to doom us.

Yet there is no sign that we, as a species, are about to stop insisting that we can continue to do business as usual forever.

Thoughts like this make me happy that I am too old to expect to see the coming disaster. However, I do have a daughter, a son-in-law, and grandchildren, as well as nieces, nephews, and now even a grand-nephew. I hope they will all be among the survivors.

Limiting Factors

When I discussed the question of whether
there are limits to growth. I noted that endless
growth—in population, in fossil fuel use, and even
in the global economy—is not possible and that
those who insist that it is are guilty of crimes
against humanity.

Now I would like to introduce the ecological
concept of *limiting factors*. Simply put, a
population can grow only until it uses up the
resource in least supply. That resource is the
limiting factor.

Consider a small car factory. It has fifty car
bodies, three dozen chassis, eighty engines, and
twelve wheels, with tires. How many cars can it
make?

Three, if each one has four wheels. With a
little reengineering, you could get four three-
wheeled cars. All the bodies, chassis, and engines
beyond that are of no use. Wheels are the limiting
factor.

The same line of thought applies to
populations of humans, other animals, and
plants. Since we are chiefly concerned with
humans, we can ignore other living things (though
they must obey the same rules).

Before we go further, what are the human
equivalents of car bodies, chassis, engines, and
wheels? We need living space, food, water, and air.
Beyond those basics, there is a long list of things
we would miss so badly if civilization collapsed
(how about toilet paper?) that we might consider
them needs as well.

But back to the basics. Living space is not a
limiting factor. As is often pointed out by people
opposed to the idea that there is such a thing as
overpopulation, the entire global stock of humans
could be put in Texas, with room left over. Of

course, that doesn't count the farms needed to feed the crowd, or the forests needed to supply the TP.

Air—or the oxygen in it—isn't a limiting factor either. There is much more than we need, at least as long as we don't kill off the green plants and ocean algae that replenish the oxygen for us.

Food, on the other hand, is definitely a limiting factor. Russell Hopfenberg points out that "Human Carrying Capacity Is Determined by Food Availability," *Population & Environment* (November 2003). That is, you can't grow more people than you can feed.

Growing food requires soil and the nutrients in the soil, often supplied in the form of fertilizer. One component of fertilizer is phosphate, obtained by mining a diminishing supply of phosphate-containing rock. When we run out of that rock, we'll have a serious problem.[1]

Components of food such as vitamins and minerals are also important, but shortages of these can be made up, at least as long as we can make vitamin pills.

Water—fresh water—comes in because although we need some for drinking and washing, we need a great deal more for irrigating food crops. If we don't have enough water, we don't have enough food. And unfortunately, global climate change (or global warming) is projected to increase the frequency of droughts and heat waves, as well as to diminish snow in mountains such as the Sierra Nevadas and even the Himalayas, as well as to make that snow melt earlier in the spring, which will mean less water available during the summer growing season.

But surely if less water runs down the rivers, farmers can just pump more out of the ground? The trouble is that they've been doing that for decades already. Water tables are dropping, so that it is harder to pump water to the surface. Some aquifers are nearing exhaustion.[2]

Despite the looming phosphate problem, it seems reasonable to say that the ultimate limiting factor for human beings is water.[3] Indeed, a quarter of the world's population currently face the prospect of severe shortages. "Water stress is the biggest crisis no one is talking about. Its consequences are in plain sight in the form of food insecurity, conflict and migration, and financial instability." says Andrew Steer, president and CEO of the World Resources Institute.[4]

There is, of course, a huge supply of salt water in the oceans. It's possible to remove the salt from, or desalinate, sea water, but that is a very energy-intensive process. If we become dependent on that, energy supply may be the ultimate limiting factor. There is also a lot of fresh water locked up in glaciers and especially in the ice that covers Antarctica. There have been proposals, and pilot projects, to tow icebergs to places that need water and let them melt. The latest such proposal comes from South Africa.[5] But that possibility is only good until global warming melts all the ice. That day is centuries off, we think, but...

"Carrying capacity" is another crucial ecological concept. It is defined very simply as the size of the population that the environment can support, or "carry," indefinitely, through both good years and bad.[6] It is not the size of the population that can prosper in good times alone, for such a large population must suffer catastrophically when droughts, floods, or blights arrive or the climate warms or cools. Of course, *really* bad times do happen, and both local populations and species can go extinct as a result. But you want to put that time off as long as possible.

Carrying capacity is a long-term concept, where "long-term" means not decades or generations, nor even centuries, but millennia or more. It is thus in alarming contrast to what human society's decision-makers think of as long-term—at most, five or ten years.

What is Earth's carrying capacity for human beings? It is surely impossible to set a precise figure on the number of human beings the world can support for the long run. As Joel E. Cohen discusses in *How Many People Can the Earth Support?* (W. W. Norton, 1996), estimates of Earth's carrying capacity range from under a billion to over a trillion. The precise number depends on our choices of diet, standard of living, level of technology, willingness to share with others at home and abroad, and desire for an intact physical, chemical, and biological environment (including wildlife and natural environments), as well as on whether or not our morality permits restraint in reproduction and our political or religious ideology permits educating and empowering women. The key, Cohen stresses, is human choice.

It is worth stressing that we get the largest carrying capacity only with minimal per capita use of resources (or minimal standard of living). Think of living in a grass shack and burning dung to cook your daily rice or beans. No doctors or hospitals or supermarkets or cars or electricity or Internet or... And then we are subject to disease and famine as we have not been for the last couple of centuries.

If you want a higher standard of living—something like what you are used to today—you will need a much smaller population. If you want that standard of living for everyone, you may need a world population of less than a billion.

Getting there is the problem. I doubt very much that we can do it voluntarily. But running

down our fresh water and phosphate supplies
might do the job for us.

What's for Dinner?

It's not news that the size of the human population—7.6 billion and counting[1]—poses problems. One of the major problems is that of how to feed everyone.

Sure, the supermarkets are still full, at least in the developed world. But people are recognizing that raising the huge amounts of meat we consume is not really sustainable.[2] For one thing, it releases large amounts of carbon dioxide into the air and thus contributes to global climate change (or global warming). Perhaps more to the point, it takes an enormous amount of land to raise the grain used to feed the pigs, cattle, and chickens whose meat, milk, and eggs we relish. And both of those factors will get worse—the world's human population is expected to hit 11 billion before the end of the century.

There are things we can do. One depends on the way food chains work.

"Food chain" is a simple concept. A chain begins with those living things that convert solar (or sometimes chemical) energy and simple chemicals such as carbon dioxide into organic material. We're talking plants that use sunlight to make leaves and stems and roots and fruit.

The next step in the chain is animals that eat plants (herbivores). The step after that is animals that eat animals that eat plants (carnivores). And then there are the animals that eat animals that eat animals that eat plants. And so on.

It's more complicated than that, of course. We humans take our food from all levels—plants, herbivores, carnivores—of the food chain. Even herbivorous cows have been known to stomp squirrels and eat them. Ticks suck blood from both herbivores and carnivores. A more realistic depiction of what-eats-what is called a food *web*.

Just to keep things simple, though, let's stick with the food chain. If you have 100 pounds of corn, that's enough to grow about 10 pounds of cow (or cow product—milk and cheese), which in turn is enough to grow about 1 pound of cow-eater (people, for instance). The rest goes to running the cow. Some herbivores (chickens, for instance) are more efficient than that at turning feed into meat, but by and large you lose about 90 percent of the food value at each step in the food chain.

Imagine a human population that eats nothing but meat. If it shifts to eating nothing but plants such as tomatoes, beans, potatoes, and grain, that diet will support a population ten times as large.

NO. That does not mean we can feed a population of 75 billion by just shifting to a vegetarian diet. Much of the world's population already eats very little meat (they can't afford it). Even in the developed world, people don't eat 100 percent meat.

Think about it. What did you have for dinner last night? Even if you had pizza, most of what was on your plate was crust and sauce, not cheese and pepperoni. A steak? It came with salad and potatoes. So dinner was less than half meat. If that ratio holds generally, shifting to an all-plant diet only permits you to multiply half the food supply by ten. And that is only true in the developed world, which contains only about a fifth of the total world population. So cutting out meat entirely might make it possible to boost world population to 15-20 billion.

People are already developing fake meats more convincing than tofurkey (have you tried the Impossible Burger[3] yet?) Are we going to be stuck with that? Do we really have to give up meat entirely? Surely not, for some meat will still be raised for those who can afford high prices. And then there's lab-grown meat[4]—take a few cells from an animal, convert them to muscle cells, let them multiply in a vat, harvest them, and shape the resulting mush into patties. They're working on improving the texture and bringing the price down (waay! down). In time... And then it gets weird, for the cells can come from any animal at all, including humans. Just think of what will happen when the celebrity marketers get ahold of this! Your grandchildren could dine on very special rump steaks.

What about fish? There are already projections that wild-caught fish will be gone by 2050.[5] We'll be almost completely dependent on farmed salmon, shrimp, tilapia, carp, and catfish. Researchers are also working on ways to farm tuna and cod, and perhaps others. Unfortunately, farmed fish need to be fed, and most of their feed is made from other, smaller fish. The vacuuming of the world's oceans will not end when the commercial-sized tuna, cod, salmon, striper, haddock, hake, and halibut (etc.) are gone. It will end only when there are too few small fish left to be worth hunting for them.

At that point we will need to put the farmed fish on a vegetarian diet. That may mean that we will shift to less fussy carp and catfish. Yet even those may not serve us long, for one recent study suggests that global warming may make life difficult for fish farms.[6] So perhaps we'll be putting fish cells in the vats too and churning our filets out of the meat factories. Or faking the filets out of veggies—yes, they're working on that too.[7]

What about eating bugs? No, I don't mean riding a motorcycle and licking them off your teeth. People are talking seriously about how efficiently insects convert plant material to protein and how "green" a bug-based diet could be.[8]

However, you won't be getting your meals by waving a butterfly net through the air, for there don't seem to be as many bugs around as there used to be.[9] You may have noticed this yourself, for your windshield is probably not as bug-splattered as it used to be (the English call this the "windscreen phenomenon"[10]). So we'll be seeing either bug-farms or bug-vats.

Anything else? Well, the Diet of Worms was a 1521 attempt to figure out what to do about Martin Luther. But people *do* eat worms[11] ...

Okay. Yuck. People do just fine on a vegetarian diet. It's healthier, more sustainable, greener... And it can be faked up to look, taste, and feel like meat. Or so they say.

So what kind of veggies will we be eating? The obvious answer is that they will be genetically engineered or modified. The very idea offends some people, for GMO crops are not what nature made. On the other hand, neither are conventional crops, which have been drastically altered from their "natural" form by centuries of selective breeding. Broccoli, cauliflower, kale, cabbage, Brussel's sprouts, and other "cole crops" are all derived from a single Mediterranean wild cabbage species.[12]

People also worry that GMO foods may not be safe (though there is no evidence of problems) or that they may lead to corporate domination of the food supply (admittedly more of an issue). Proponents claim that GMO crops need less in the way of chemicals. This is true for pesticides. But herbicide-resistant crops actually mean more herbicides can be used to kill weeds. Another potential benefit is resistance to fungal and viral diseases of crops; those diseases can greatly reduce harvests.[13] Researchers are also working on giving non-leguminous crops (such as grains) the ability to make their own nitrogen fertilizer, just as legumes (such as peas and beans) already do.[14]

So far the genetic engineering or gengineering of crops has barely begun. Someday potatoes may taste like steak or lobster. Perhaps there will be entirely new types of veggies, or weird hybrids. Or perhaps there will come a time when tomatoes and corn (etc.) can grow on trees. That sounds far-fetched, but orchards don't need herbicides (you just mow the weeds between the trees), soil erosion is not a problem, and deep roots may find more water. Trees also don't need to be replanted every year, and every year, they get bigger and bear more fruit.

What's for dinner? For the next few years, same old, same old. But as we approach 2100, we'll be seeing less "real" meat on our plates and more lab-grown meat and plant-based fake meat. Insects and worms may have some niche appeal. Once we pass 2100, we'll start seeing some weirdly different gengineered fruits and vegetables.

The next question is how many hands will be reaching for the pot in the center of the table. The answer depends on how well we deal with other problems, such as global warming.

Bouncing, Bouncing, Bouncing

Sustainability—defined as using resources in such a way that resources remain for future generations to use, with no decline in life-style or well-being—entered the global debate in the 1980s, when the United Nations' World Commission on Environment and Development released *Our Common Future* (Oxford University Press, 1987) (the Brundtland Report). It recognized that limits on population size and resource use exist and must be taken into account when governments, corporations, and individuals plan for the future. The 1992 Rio Conference established that sustainability implies such things as cutting forests no faster than they can grow back, using ground water no faster than it is recharged by precipitation, emitting no more pollutants than the environment can absorb or destroy, stressing renewable energy sources rather than exhaustible fossil fuels, and farming in such a way that soil fertility does not decline.

Some still cling to the idea that a sustainable civilization is achievable, if we can just reduce both population and consumption, or—lacking that—if we develop suitable technologies. Despite the title of his article, Andrew Winston ("The Story of Sustainability in 2018: 'We Have about 12 Years Left,'" *Harvard Business Review*, December 27, 2018) is optimistic about changes in technology and corporate practice.

But population has continued to increase, as has resource use, including the exhaustion of groundwater aquifers and the use of fossil fuels and concomitant carbon emissions, which threaten a warmer climate. Others have given up on sustainability, replacing it with "resilience"; Melinda Harm Benson and Robin Kundis Craig[1] argue that we must focus on coping or adapting instead of sustaining.

Even in a sustainable world, we must cope with and adapt to the results of what that world throws at us. We shift from "sustainable" to "resilient" when we start seeing those adverse events as coming more frequently and being more extreme. Ordinary coping and adapting are no longer enough. The concept of resilience recognizes that we cannot maintain a steady state such as sustainability envisions. It implies that conditions will get worse but we will struggle to restore normality, to recover or bounce back from events such as the droughts, heat waves, wildfires, floods, hurricanes, and other consequences of global warming, exhaustion of groundwater, shortfalls in food supply, and the death of wild-caught fisheries, as well as from more random events such as volcanic eruptions, earthquakes, and tsunamis. In effect, "We know we are going to get hammered; the question now is whether we can bounce back."

Is that what we will have to worry about for the next few decades or even centuries? Bouncing back? Disaster recovery? The random events are always lurking in the background, essentially unpredictable. The consequences of population growth, resource use, and carbon emissions are definitely predictable. They are thus, concerns for the present and the immediate future.

What will we have to bounce back from? Well, one of those consequences is exhaustion of groundwater, essential for agriculture and renewable only on a time scale of millennia. Recovery must mean finding new sources of water, such as by desalinating seawater or towing icebergs from Antarctica (both expensive processes). Another consequence is bigger hurricanes, which destroy both individual houses and whole towns. So do wildfires and floods. Recovery means reconstruction (which is good for the economy—all those construction jobs as well as purchases of bricks and lumber), and that is straightforward enough. Droughts may mean widespread famine and death. Rising seas, particularly in low-lying lands such as Bangladesh, will mean displaced populations and increased conflict—and deaths. For deaths, recovery means replacing the dead with new births.

And it's all good if the world gives you time to rebuild or rebreed before the next catastrophe hits. But what the experts are saying about the consequences of global warming is that we can't expect that. We are going to get hammered again and again and again, probably in rapid succession. There may even be near-simultaneous hits, as California found in 2018. Drought parches the vegetation, leading to wildfires that remove the vegetation that holds the soil in place. And then come the rains and the mudslides. In 2020, we had the Covid-19 pandemic, which killed many and cost many, many more their livelihoods for months on end. On April 21, 2020, the UN issued a forecast of widespread "famines of biblical proportions" due to the combination of the pandemic, plagues of locusts in Africa, and global climate change.[2]

The trouble with resilience... It is a property of rubber balls. Drop one and it bounces back—but not all the way to your hand. It falls slightly short. If you let it bounce-bounce-bounce, the bounces get shorter and shorter until the ball just quivers and rolls away into the corner of the room.

If resilience-instead-of-sustainability works in even a remotely similar way, we are not going to be able to restore normality completely after every adverse event. We will fall short at least some of the time, and then even shorter. We are looking at continual decline in population, built infrastructure, resource use, food production, and so on. Bounce, bounce, bounce, and dribble to a much reduced steady state in the corner of the room.

Am I just catastrophizing? In 2009, Hans Joachim Schellnhuber, director of the Potsdam Institute for Climate Impact Research, told the Copenhagen climate conference that if global warming passes 5 degrees Celsius (9 degrees Fahrenheit) warmer than today, by no means a worst-case scenario and well below maximum projections, the Earth may become able to support no more than one billion people, a number last seen in 1800.[3]

It won't happen right away. That bouncing ball takes time to run down. But it's something to worry about for your grandchildren or great-grandchildren.

So what can we do about it? No one is willing to talk about reducing population. No one is willing to reduce carbon emissions enough and some are still insisting they aren't the problem. Some people are talking about geoengineering—spraying sulfates or water droplets into the stratosphere to reflect some of the incoming solar energy—but that idea is untested, the side effects are not well understood, and once you start and have it working, you don't dare stop. (Unfortunately, politicians have a way of pulling the plug once the problem is no longer apparent. U.S. Republicans have actually tried to roll back the Clean Air and Water Acts because the air and water are clean now. Pollution is gone, so we no longer need to control it. Right?)

It might work better if we could just pull excess carbon dioxide out of the air, as described in Richard Conniff, "The Last Resort," *Scientific*

American, January 2019. A number of techniques are being developed and even tested, but to work they need to be deployed within the next few years. And they will be very, very expensive. By 2100, we will need to pull a trillion tons of carbon dioxide out of the air, at a cost of $100 or more per ton. The U.S. Gross National Product is about $18 trillion.

It would be cheaper to shift away from eating meat and plant the rangelands no longer needed to grow cows with lots and lots of trees (which unfortunately need more water than grass does). That would help, though only temporarily if we keep growing the population and burning fossil fuels.

Unfortunately, thinking in terms of temporary help is short-term thinking. We need that, when the problems we are trying to solve are urgent. But we also need to look further into the future. We need solutions that last.

Of course, the solutions we need are inherent in the problems we face: Stop doing that!

I wish I had a better one, one that politicians could accept.

Gee, It's Lonely Here

We don't have to worry about this one at the moment, but eventually we will surely join the thousands of civilizations that have preceded us in the Milky Way galaxy. Not that we seem to have any chance of ever meeting them.

Seventy years ago, famous physicist Enrico Fermi was having lunch with colleagues (including Edward Teller of H-bomb fame). Conversation wandered, as conversation does, until Fermi burst out with, "Where are they?"

What he meant was, "Where are the aliens?" He then backed up the question with some quick calculations on the probabilities of planets like Earth, of life, and of high technology and concluded that we should have been visited many times. So, "Where are they?"

This is the Fermi Paradox, and if you believe in UFOs and alien abductions and other such myths, all the way back to Ezekiel's wheel, it's no mystery. We have been visited, many times, and aliens even live amongst us, as well as in caverns underground and on the dark side of the moon. But, Fermi noted, there is no evidence. Only folklore and fantasy.

So. The Fermi Paradox. It has puzzled many people over the decades since that lunch. In *Where Is Everybody? Fifty Solutions to the Fermi Paradox and the Problem of Extraterrestrial Life* (Copernicus Books, 2002), Stephen Webb described Fermi and his paradox and offered a variety of answers that have been suggested—most seriously, some a bit tongue-in-cheek—for why the search for extraterrestrial intelligence (SETI) has not succeeded. John Gribbin, in "Alone in the Milky Way," *Scientific American,* September 2018, argues that the conditions that led to our technological civilization would be difficult to match even among the 100 billion other stars of our galaxy.

By 1961, radio astronomer Frank Drake had already been listening for radio signals from other civilizations for 4 years.[1] In that year, he formulated what came to be known as the Drake Equation to calculate N, the number of civilizations in the Milky Way galaxy whose electromagnetic emissions are detectable at the present time:

$$N = R_* \times f_p \times n_e \times f_l \times f_i \times f_c \times L$$
where:

R_* = the rate of formation of stars suitable for the development of intelligent life;

f_p = the fraction of those stars with planetary systems;

n_e = the number of planets, per solar system, with an environment suitable for life;

f_l = the fraction of suitable planets on which life actually appears;

f_i = the fraction of life bearing planets on which intelligent life emerges;

f_c = the fraction of civilizations that develop a technology that releases detectable signs of their existence into space;

L = average length of time such civilizations release detectable signs of their existence into space.

This equation has impressed many people whose eyes widen at the sight of anything that looks mathematical. But it was never intended to be a way to calculate how many ETs are out there waiting for contact. Rather it was a way of defining our ignorance, of saying "This is what we need to know before we can estimate." At the time, we really knew only one of the terms, R*, the rate of formation of suitable new stars (about 10 per year[2])

. Since then, we've found enough stars with
planets that we can venture a reasonable guess at
fp, the fraction of those stars with planetary
systems. But all the rest of the numbers? We have
no idea, for we know of only one place—Earth—
that has developed life, intelligence, and
technology. That is too small a sample to work
with.

What about **L**, the average length of time
technological civilizations release detectable signs
of their existence into space? In the 1960s, some
people put that as low as 100 years. The estimate
grew as the Cold War waned and the threat of
nuclear annihilation receded. With the advent of
communications satellites and fiber optic cable,
we realized that even if a civilization survives,
technological change may reduce or eliminate
leakage of radio signals into space, so the estimate
shrank again. Today, with political chaos
threatening to engulf the planet, it should perhaps
shrink some more, perhaps even back toward the
Cold War number. And of course a civilization
does not have to end in Armageddon. We have
problems of overpopulation, global warming, water
supply, other resource shortages, and more that
could do the job. And then there are asteroid
impacts, solar flares,[3] and other catastrophes,
only a few of which we can hope to do much
about. Planetary civilizations—or ones with
detectable technologies—are very unlikely to last
forever.

A value of **L** in the range of centuries rather than
millennia or eons does not seem unreasonable.

So what does the Drake Equation look like if we put
in some numbers?

$$N = R_* \times f_p \times n_e \times f_l \times f_i \times f_c \times L$$

Where:

R_* = 10 stars suitable for the development of intelligent life per year;
f_p = .5 of those stars with planetary systems;
n_e = .1 planet, per solar system, with an environment suitable for life;
f_l = 1 suitable planet on which life actually appears;[4]
f_i = .1 life-bearing planet on which intelligent life emerges;

f_c = .1 civilization that develop a technology that releases detectable signs of their existence into space;

L = 100 years during which such civilizations release detectable signs of their existence into space.

N = 10 x .5 x .1 x 1 x .1 x .1 x 100 = .5 civilizations available to contact in the Milky Way galaxy at this moment in time. [5]

We're here, now, so it rather looks like we're alone (although there may well be technological civilizations that do not use radio technology). There may have been many in the past. There may be many in the future. But very few will ever coincide in time. We need to bear in mind that cosmological time is vast.

You can fiddle the numbers all you want. Go ahead and make **L** 10,000 years instead of 100. N becomes 50, which is still not impressive to anyone raised on Star Trek and Star Wars. An **L** of 10,000,000 gives you an **N** of 50,000. An **L** of 10,000,000,000 gives you an **N** of 50,000,000, which would surely make the Trekkies happy. But remember—you're playing a game largely of optimistic ignorance.

So are we alone, at least for now?

If not, where are they?

If so, can we somehow stretch out our own **L** until we have a chance to find someone?

We can of course do our very best to deal with resource shortages and political upheavals and ward off asteroid impacts, etc. But Mother Nature has a great many ways to produce disaster, and extinction is normal (99 percent of all the species that ever lived on Earth are extinct). Our best chance for long-term survival is probably to widen the target zone by getting off the planet. That means space colonies, either in habitats (giant space stations) or on other worlds. Since it seems impossible to travel faster than light, we are not going to be able to go very far outside our solar system—not past a few of the most nearby stars. If the colonies in turn spawn colonies, humanity could spread out into the galaxy, albeit very slowly. So could ETs, of course, so their absence again suggests they aren't there.

The apparent impossibility of faster than light travel, by the way, also speaks to the lack of ET visitors. If they can't get here from there, we won't see visitors even if the galaxy has as many ETs as a fig has seeds. In that case, though, we should be picking up signals. Since we aren't...

Meanwhile, is it worth our while to continue listening for extraterrestrial radio signals, still the major component of SETI? As science projects go, it's relatively inexpensive and the potential payoff is large, for we could learn a great deal from another civilization, Even its mere existence would be informative. There is also the point that, even if we are alone in the galaxy at this time, there may well have been technological civilizations in the past whose radio signals are still spreading through space. Space is vast, and stars are so far apart that radio signals can take thousands of years to pass from star to star. If we ever detect such radio fossils, SETI will become an exercise in xenoarcheology.

But face it—the odds seem slim.

SOME IMPLICATIONS OF
GLOBAL WARMING

It is hard to avoid mentioning the threat of global warming when discussing the constraints we face in the not too distant future. That threat also deserves attention in its own right, especially considering recent news reports from Europe, India, and Australia.

The threat is not just a matter of droughts, heat waves, and rising seas. Those events are going to drive millions of people from their homes. When these people come up against borders... well, the International Panel on Climate Change (IPCC) has been predicting for years that global warming will lead to conflict. And President Trump has pulled the US out of an important arms control treaty with Russia.

It is enough to make one twitchy. When I was a kid outside Washington, D.C., getting duck-and-cover drills at school (we might as well have watched the show from the playground—we couldn't have died any faster), I thought it would be a miracle if I lived to see the year 2000. The Cold War ended (thank goodness!), but now I'm wondering if we are facing a very similar threat.

The Cassandra Code

In a recent Facebook thread, I was opining on the report that "Climate change could pose 'existential threat' by 2050,"[1] saying that this really underlines what some people are saying—we have ten years to get our asses in gear. Maybe less than that. But if you read the report, a lot of it is about how climate problems (heat, drought, high seas) are going to put billions of people on the move away from home. There will be massive security concerns. I expect Trump's call for a wall will seem a whisper in a hurricane. There will be many deaths from climate effects, but even more as nations defend their borders. My daughter remarked that "I am convinced this is correct. Giving voice to it, makes one a Cassandra immediately. Most people just don't know where to put that."

The Cassandras have never been well received. Yet today, the developing climate crisis has spawned a considerable number of studies and reports, all delivering the same message that as we sail this bay in the sea of time, there are reefs ahead.

The warnings of doom go back to the 1860s, picked up steam in the mid-1900s, and have become almost commonplace today. There are two main reasons for the proliferation of doom-crying. First, our understanding of how the world actually works (science!) has increased tremendously. Second, we have developed techniques to express that understanding in ways that permit us to look some distance into the future. These techniques depend utterly on the computer and our ability to construct computer programs that incorporate what we understand of how the world works and then turn those programs loose upon the vast amounts of data we have been collecting. The result is an extension of historical trends (visible in all that data) into the future. The accuracy of the extension does not depend on whether we are liberals or conservatives, or Christians, Moslems, or atheists. Rather, it depends on how well we understand the world and is therefore subject to correction or improvement as our understanding improves. It also improves as our computers and computer programs get better.

These programs are computer simulations. Since the nuclear industry refers to such simulations as "computer codes," the title of this essay is "The Cassandra Code."

The original Cassandra was a daughter of King Priam and Queen Hecuba of Troy. She was cursed to utter prophecies that would come true but that no one would believe. The modern Cassandra codes are not quite so cursed, though they are often maligned by critics, who say such things as "They're just computer programs! They can't know the future!" and "They just say what they're programmed to say! Garbage in, garbage out!" and "But you have to oversimplify too much to put it in the computer!"

Computer simulations attracted considerable attention during the 2020 Covid-19 pandemic. They predicted that large numbers of people would die. When those numbers seemed too large, some people criticized the models as being too pessimistic. But think about it. If you can go shopping only every other week, a computer model of your consumption habits might say you need six loaves of bread. So you buy six. If you really only need four, that's okay. You're covered. If the model had said you needed only two, and you really needed four, the model would warrant criticism.

In other words, a pessimistic model is the kind you need. If the model is too optimistic, and you are trying to evaluate something more serious than how much bread you need, such as how serious a pandemic is and how many ventilators you need, you are screwed.

Computer simulations have scored some notable successes, as when they were used to simulate the effects of putting megatons of crap in the air (as by nuclear weapons). The result is a rapid chilling of the global climate, which gave rise to the term "nuclear winter" and made it clear that a nuclear war involving as few as a hundred warheads could have no winner. This led to the end of the Cold War and the threat of nuclear Armageddon under which I grew up.

They scored another win when they showed that adding chlorofluorocarbons (CFCs) to the atmosphere was destroying stratospheric ozone, creating the "ozone hole" scientists had observed, and increasing the risks of skin cancer, cataracts, and other ills. They also said that if we could stop doing that, the ozone hole would begin to recover and the risks decline, and once the world's nations signed the Montreal Protocol and banned CFCs, that's what happened. (Although recent reports reveal that factories in China are pumping CFCs into the air again.)

Cassandra codes also sparked a long debate over the future of human civilization when, in 1972, *The Limits to Growth* was published. The point of the book was that trends visible in the data as of 1972, projected into the future by means of a computer simulation that was ludicrously simple by today's standards, led to serious trouble in this century. To prevent that trouble would require reducing pollution (including carbon dioxide emissions), industrial development, and even population.

No one liked that news, and the book was vigorously attacked. But follow-up studies, each using decades' more data and improved computer simulations, have repeated the message over and over again. We have done nothing to change the trends. Therefore, we have done nothing to ward off the trouble the simulations project. Today that trouble looks a lot like the existential impact of global warming.

A critical question is "Why have we done nothing to ward off the trouble the simulations project?" If we put it another way, it becomes "Why could the Cassandra codes succeed in the cases of nuclear weapons and CFCs but not in the case of global warming?"

I put it to you that most national and military leaders did not really want nuclear weapons (or at least nuclear war). As World War II showed, they were overkill. But the logic of mutual assured destruction (MAD) meant that if your rival had them, you had to have them too. The existential threat of nuclear winter gave everyone an excuse to back off and then sigh in relief.

People wanted CFCs. They were in refrigerators and air conditioners. They put the Pssstt! in aerosol cans. They had industrial applications. But they were not essential to anyone—alternatives existed. It was relatively easy for everyone to agree to do the right thing.

Global warming seems to be a different matter. Since it is due in large part to increasing amounts of carbon emissions, largely from burning fossil fuels (coal, oil, and natural gas) and clearing of land (removing trees that pull carbon dioxide out of the air), we have known for decades what to do: STOP DOING THAT!

But we need the energy we get by burning fossil fuels. We need electricity for our lights, air conditioners, and computers. We need our cars. We love central heating. And people make huge amounts of money from doing things the ways we always have. The energy companies and the politicians they bankroll will give up fossil fuels only over their dead bodies.

Not that we would suffer if that happened. We have alternative energy technologies such as wind, solar, tidal, and geothermal. And the companies behind them are trying hard to become big enough to replace the big fossil fuel companies. But it is hard to make progress against entrenched giants.

We may need that existential threat of global warming. It doesn't threaten a "nuclear winter," but rather a hot "carbon summer." If the message reaches enough people, the result may be similar to the end of the Cold War.

Unfortunately, I suspect that the carbon summer will have to become real, not just the anguished cry of a Cassandra code. By then, in another decade or so, it may well be too late.

Just as it was when the people of Troy ignored Cassandra's warning that they really, really shouldn't haul that big wooden horse inside the walls.

Just A-Looking for a Home

Summer 2019 news was depressing. A break-all-records heat wave in Europe that moved to Greenland and melted enough ice to add measurably to global sea levels.[1] A drought in India that has people saying forty percent of the country's population may have no access to drinking water by 2030, and parts of India may get too hot for humans to live there.[2] By the end of the year, Australia was beginning its summer with droughts, record heat waves (pushing 120 Fahrenheit), and bush fires that beggared description.[3]

There's some hype in that. The media love bad news. In India, reports said the monsoon had arrived for this year, although it was not really enough; there was still a "monsoon deficit." But even if the situation eases next year, or the year after, climate change is real. The droughts and heat waves will return, and they will get worse. This is just the first taste of the "carbon summer."

There are solutions for India, ranging from getting serious about water conservation to desalination of seawater to towing icebergs from Antarctica. All are expensive, and the need for freshwater is huge—India has a population of 1.339 billion (2017), a significant fraction of world population.

Will such solutions be enough? We can hope so, but if they are not, a large portion of India's population will face a stark choice: die or move.

Fortunately, some parts of India are still wet. Many people will be able to move within the country. Thousands already are, to the point where villages are being left empty except for the elderly and the ill who cannot walk away. But the destination zones are already occupied. There surely is not room for everyone.

On the other hand, the Indian government has now erased special protections for cooler and fertile Kashmir,[4] which has enjoyed semi-autonomous status and been allowed to keep the rest of India out for the last 70 years (ever since the post-British-occupation partition). The ostensible reasons are to put all of India under the same rules and to stimulate economic development. The real reason may be that India's government is eyeing the upcoming "move or die" moment, thinking internal migration will be better than external, and reducing any barriers that might get in the way.

After all, where else can they go?

If the U.S. were facing similar problems, President Trump would surely be talking about annexing Canada. And Canada does not have enough of a military to do much about such a move.

Nor does Denmark, which I mention because Trump has been talking about buying Greenland.[5] Once the ice melts, which may not be too far off, it will offer a sizable amount of land with a temperate climate, a good place to move people from areas plagued by heat waves and droughts. There's no soil under the ice that's there now, but once the ice is gone there will be plenty of space for greenhouses and homes. However, I do not think current American politicians are capable of that sort of long-term thinking.

Do the Indians have neighbors they might annex? There's Kashmir—Oh, they just did that. There's Pakistan. It's already occupied, of course, and it's well armed, even with nukes. Southeast Asia is geographically smaller and closer, just across the Bay of Bengal. But it too is already occupied, and it too is well armed. Yet these folks too may be looking for new homes. It's not just India. The news is already asking "What Happens

Some Implications of Global Warming

When Parts of South Asia Become Unlivable? The Climate Crisis Is Already Displacing Millions."[6]

China? Fuhgedaboutit.

Africa? Despite the continent's own problems, India might well see it as offering an answer. It's certainly spacious enough, parts of it are still wet, and I don't think its collective militaries could do more than slow down an invasion. And an invasion it would be. India has enough shipping to haul large numbers of troops and refugees across the Arabian Sea and Indian Ocean very quickly.

The next question is how the rest of the world would react if they tried it. India's chief opponents are Pakistan (already rattling its sabers about Kashmir) and China; its allies include Iran and Japan, and at least sometimes the United States[7]— which, by the way, is withdrawing from arms control treaties and developing new missiles.[8]

Who would defend Africa? Many European nations have historical and economic interests there, and the EU and the African Union have a partnership agreement.[9]

That is quite a lineup for a World War III[10] fought in India, Africa, and the Indian Ocean. Leaders in the US, Europe, Russia, and China would surely prefer that for a venue. So would white supremacists. If it came to that, the death count would probably be considerably greater than if the Indians just stayed home.

We could help. As I noted above, there are potential solutions to the lack of water. For instance, solar heat can be used to power desalination, and the U.S. Department of Energy is funding research into better methods.[11] Such methods would be helpful not only in India.

Heat can be controlled indoors if there is electricity for air conditioning. That takes money. Outdoors is another matter, so it would be crucial to move crops and livestock indoors as much as possible. Greenhouses and barns are easy to build. But again, money.

And money. And more money.

We should have started spending on preventing global warming years ago, when it was obviously on its way.

But even if we have to spend trillions helping India survive, it will be better than spending trillions on World War III.

Note that India is by no means the only place whose people may be looking for new homes in the not-too-distant future. We're already seeing people fleeing Africa and the Middle East for Europe, and Central America for North America. What they're fleeing is mostly political chaos, some of which reflects environmental disaster. But the latter is going to play an ever bigger and more direct role. South Asia will be facing more heat and floods. Australia heat and drought and more fires. Rising seas doom many island states and will affect every nation with a coastline. And unfortunately, according to the UN's Refugee Convention "Climate migrants are not legally considered refugees according to international refugee law."[12] They are therefore not entitled to protection, and borders may be sealed against them.

That old Boll Weevil Song[13] is going to gain new life.

The Problem of Sunk Costs

It is disappointing that the December 2019 United Nations Climate Summit (COP 25) failed to do much about global climate change other than declare a need for nations to be more ambitious about reducing carbon emissions and helping poorer nations cope with the impacts of global warming. A discussion of global carbon markets was put off till the 2020 Summit.

We know that many parts of the world face an existential crisis. Heat waves are already beyond all precedent. Sea level rise will surpass two feet by 2100. Billions of people will be affected, even in the richer nations of the world.

So why aren't we doing anything about it?

We talk about denialism, about politicians refusing to believe scientific analyses, perhaps at the behest of the billionaires who fund their campaigns. We talk about corporate interests—not just fossil fuel companies, but also car makers, cement makers (and users), and highway builders—that want to keep on making money just as they always have. Business as usual, and that's the scenario in all the climate models that leads inexorably to disaster.

We shake our heads. We rail at their short-sightedness. We call them stupid and blind and evil.

But... And I don't mean to excuse their behavior, but to offer an explanation. According to economist Frank Ackerman, "The ultimate economic obstacle to climate policy is the long life of so many investments."[1] The fossil fuel industry has a ginormous amount of money—on the order of ten trillion dollars—invested in coal mines, oil and gas wells, tankers, refineries, pipelines, gas stations, and more. That money is called sunk costs. It's already been spent. It can't be recovered. It can't even be repurposed.

To eliminate carbon emissions means, among other things, to stop using fossil fuels as soon as possible. That means turning sunk costs into "stranded costs," which are no longer serving any useful purpose. It means writing them off. It means turning a profitable industry into a massively unprofitable one, to the great dismay of its investors.

It should be no surprise that corporate interests are reluctant to bite the climate change bullet, no matter how necessary doing so may be.

Can we *make* them bite that bullet? The past may be instructive. By the 1940s, the developed world had a massive industrial infrastructure well suited to the needs of the time. World War II saw the United States' portion of that retooled in a hurry to produce war materiel. Europe and Japan saw their portions destroyed and were thereby freed from the problem of sunk costs. When they rebuilt after the war (with U.S. help such as the Marshall Plan), the U.S. found itself at a competitive disadvantage because much of U.S. manufacturing depended on older, obsolete facilities. It took a couple of decades to adjust.

I'd hate to think we need a war to make coping with climate change more feasible or practical. But we do need that level of replacement of old infrastructure.

And unfortunately, we are seeing the rise of populist demagogues who promise their people that older ways can be restored and nations can be great again. We are seeing barriers to the movement of environmental refugees. We are seeing the abandonment of treaties intended to forestall or limit the ferocity of war. We are seeing the death of alliances. We are seeing new weapons being tested and deployed.

Some Implications of Global Warming

World War III becomes easier to imagine every year.

If it happens—if we can refrain from using our nukes—the consequences may not be entirely negative.

If it doesn't happen—the U.S. has retooled before. It can do it again.

It *must* do it again. In which case—soonest begun, soonest done.

Throwing Shade

One of the scariest things about the climate change or global warming threat is that over and over again, the projections for the next few years seem to be too conservative.[1] If this continues, temperatures and sea level will go up more, and sooner, the ice in Greenland and Antarctica will melt faster, there will be more wildfires, droughts, heatwaves, storms, floods, and so on. Even the food supply will be affected. In October 2018, the Intergovernmental Panel on Climate Change (IPCC) stressed how bad things may get.[2] A year later, the IPCC said that Greenland's ice was melting so fast that we may see two feet of sea level rise by 2100.[3]

The worst hit will be developing nations and the poor. The human species is probably not in danger. Civilization will probably continue. But there will be a lot of damage to pay for, and insurance companies are going to go bust.[4]

The frustrating thing is that we have known for two generations what to do about it: Stop putting greenhouse gases, particularly carbon dioxide, into the air. Stop burning coal, oil, and natural gas. Stop clearing land, too, for cutting down a forest means putting all the carbon in the trees' wood back into the air—and the trees aren't there to keep pulling carbon *out* of the air. It's even worse if you burn the forest, as Brazil was doing in 2019.[5]

Yet decision-makers have refused to believe we have a problem. They have, in fact, shown great dedication to throwing shade. All this global warming bushwah is just a Chinese hoax, says U.S. President Donald Trump.[6] All those scientists are just trying to get rich off government grant money! And even if it were real, humans aren't to blame. Climate change is natural! Carbon dioxide is a plant nutrient not a pollutant!

Meanwhile, oil companies have buried their own research indicating a problem.[7] Admitting there was one might, after all, cut into profits. On a larger scale, trying to prevent global warming might damage the economy, and never mind that ignoring the problem will damage the economy much worse.[8]

So we're not doing much, and it's getting late to start serious efforts to prevent disaster. We may well need some serious last-ditch technology to save the day.

Fortunately, Mother Nature has given us some hints. In 1815, Mount Tambora in Indonesia blew its top. The resulting cloud of dust and sulfates reflected so much sunlight (and heat) back into space, that 1816 has been known ever since as "the year without a summer," or "eighteen hundred and froze to death."[9] In 1991, Mount Pinatubo in the Philippines repeated the lesson on a smaller scale.

The lesson is that climate can be cooled if the amount of incoming heat from the sun that is reflected back to space can be increased with some sort of "sunshade." This has led many people to think that if we could manipulate the Earth's reflectivity by geoengineering or climate engineering, we could do something useful.

Paul Crutzen suggested in 2006[10] that adding sulfur compounds to the stratosphere (as volcanoes have done) could reflect some solar energy and help relieve the problem. Roger Angel[11] suggested using a fleet of millions of small reflective spacecraft. Others have suggested spraying seawater high into the stratosphere.[12] At the time, many folks thought these guys (and others) were nuts. Today, well... The Y Combinator startup accelerator has called for proposals to turn geoengineering into businesses.[13] All approaches would be expensive, and the risk of side-effects means that such solutions should be considered only for use in extremis, if greenhouse gas reductions are not sufficient or if global warming runs out of control. Geoengineering could become our only option. If so, someone will get rich off it.

Robert B. Jackson and James Salzman[14] argue that although we need to move toward increased energy efficiency and greater use of renewable energy and explore ways to remove carbon from the atmosphere, we do also need to study the possibilities of geoengineering. "The stakes are too high for us to think that ignorance is a good policy." The need to fund the necessary research is also the subject of a two-volume National Research Council report.[15] There are signs that momentum is gathering.[16] Unfortunately, some of that momentum is coming from climate skeptics who may be preparing to embrace geoengineering as a substitute for more fundamental fixes, such as reducing carbon emissions.[17] Late in 2019, the Trump administration, which has been quite loudly dismissive of the threat, authorized $4 million for the National Oceanic and Atmospheric Administration, to assess the idea and examine proposals.[18]

Using geoengineering as a substitute for more fundamental fixes is a recipe for disaster, for if the geoengineering effort ever falters, and if carbon emissions have continued unabated, the warming effects of those emissions will kick in all at once. Imagine, if you will, bolting an air conditioner onto the side of a kitchen range. If you turn on the oven, it will get hot; this is the world without geoengineering. If you continually nudge the temperature control knob, it will get hotter; this is the world with uncontrolled carbon emissions. If you then turn on the air conditioner, while continuing to nudge the knob, the temperature will go down; this is the world with geoengineering. If you then turn the air conditioner off, the heat will come roaring back and the oven will be hotter than ever. This should be no surprise—you kept turning the temperature up, remember? Then you laid on the chill so you wouldn't feel the heat. And *then* you quit chilling.

Why on Earth would you quit chilling—or geoengineering? Politicians have a tendency to say "Problem solved! Now let's stop spending all that money!" They've actually tried to roll back the Clean Air and Clean Water Acts on that basis, not understanding that once air and water have been cleaned up, they need to be *kept* clean.

Not surprisingly, some people are extremely chary of geoengineering proposals. For instance, James R. Fleming[19] argues that climate engineers fail to consider both the risks of unintended consequences to human life and political relationships and the ethics of the human relationship to nature. They also, he says, display signs of over-confidence in technology as a solution of first resort. Most recently, the environmental group Geoengineering Monitor has issued a "Manifesto Against Geoengineering."[20] The group contends that geoengineering perpetuates the false belief that today's unjust ecologically and socially devastating industrial model of production and consumption cannot be changed and that we therefore need techno-fixes to tame its effects. The shifts and transformations we really need to face the climate crisis are fundamentally economic, political, social, and cultural.

Unfortunately, economic, political, social, and cultural changes are precisely the ones the existing economic, political, social, and cultural system resists the most. It is much simpler to implement a technological fix, even if that fix is really no more than a bandaid, a temporary measure that covers up the real problem and allows the world to go on as usual—for a while.

I'm afraid that environmental groups such as Geoengineering Monitor (and all the others that have signed the manifesto) are right when they say that climate change is not a scientific or technological problem. The science tells us, quite convincingly, that we do in fact have a problem. The technology helps us figure out what we can do about it, from renewable energy systems to geoengineering. However, much more fundamentally, climate change is a people problem. We need to change the way people around the world think, even what they value, if we are to have any real hope of warding off the worst effects of global warming.

It is not enough to think about the welfare of the wealthier citizens of the developed nations, for they have the resources to adapt to the problem. Developing nations, and the poor everywhere, do not. And they *do* matter.

It is not enough to think about the well-being of our children and grandchildren. Their children and grandchildren also matter. And theirs. Intergenerational ethics *is* a Thing.[21]

It is very much not enough to leave it at "Profit trumps all."

But... Politics.

Collapse

Discussions of global climate change or global warming often seem to turn to worst-case scenarios. The extinction of humanity doesn't seem very likely. But the collapse of civilization? That may be more worth worrying about, for there are a number of things that could make it happen. And global warming is only one of them. There are also plague, war, exhaustion of ground water for crop irrigation, Carrington events,[1] and so on.

Just what would the collapse of civilization mean? War would surely mean the outright destruction of human support systems—power plants, airports, water and sewage systems, food supply—and an immense reduction of the human population. Even without war—consider recent reports of unprecedented droughts and heat waves. In late 2019 and early 2020, Australia was experiencing record-breaking temperatures above 50 degrees Celsius (122 Fahrenheit) and extensive bush fires. The effects included loss of property and agricultural productivity. Plague removes people; when anti-plague precautions take people away from their jobs, as we saw with the 2020 Covid-19 pandemic, productivity suffers in agriculture but also in industry and commerce. The economic impact can be severe.

Heat waves and droughts don't demolish the infrastructure of civilization, but they do affect the supplies of fresh water and food and heat waves can damage power lines. If they become widespread, there could be a decline—not as abrupt as in the case of war—in the human population. Plague could have a similar impact. The first to be affected would be the poor, but if the situation got bad enough, the rest of us would be hammered too. That would in turn mean a loss of people with the skills to keep the power grid and the Internet running and the fuel supply coming for trucks and cars. Businesses that depend on commuters would crash. Cities (which hold more than half of humanity) would be in dire straits. Country folk would be in better position to survive.

If the collapse left libraries—with their troves of technical knowledge and even blueprints— intact, it might take a generation to get things running again, albeit at a reduced level. Chances are it would take several generations more to get back to being able to support a population of over 7.6 billion people. Without libraries, it would take a lot longer, though people would still have the advantage of knowing things can be done.

So how far might a collapse drop us? Bear in mind that the current world population is over 7.6 billion.

To the 1990s? The world's human population was 5.7 billion in 1995.[2] The Internet was just getting started. There were no smartphones. Young folks might consider even this much of a collapse the apocalypse.

To the 1980s? World population was 4.8 billion in 1985. We did not have the Internet yet, though email was getting started and there were bulletin-board systems. We did not have cell phones, much less smart phones. We had TV, but streaming was unheard-of.

To the 1970s? World population was 4 billion in 1975. People were beginning to talk of computer networks, though the Internet was barely a dream. The very first personal computers had arrived.

To the 1960s? World population was 3.3 billion in 1965, less than half of what it is today. We had electricity and cars. The Interstate highway system was still being built. We did not have personal computers, smartphones, or the Internet. Color TV was still new.

To the 1950s? World population was 2.7 billion in 1955. TV was new and just black and white. Computers were primitive beasts. We had electricity and cars and refrigerators. Air conditioning, though, was a rarity. It took a week to go from the U.S. to Europe by passenger ship. Air travel was still new.

To the 1920s? World population was 2 billion in 1927. We had electricity, but cars were still pretty new. Telephones were of the sort you see in old movies, with operators and party lines. For many people, refrigeration still meant putting blocks of ice in an insulated box.

To 1800? World population was 1 billion. There was no electricity, no cars, no TV, no computers, no telephones, and of course no Internet.

Even further? Collapse all the way to the Stone Age? It would take nuclear war to do that, and I hope *that* is very unlikely.

It is perhaps more pertinent to consider how rough or unbearable life would be if we were reduced to conditions equivalent to those decades. The 1990s? The screen-addicted kids who yell "OK, Boomer!" at anyone with grey hair would feel horribly deprived, but boomers and older (that's me) would find conditions pretty congenial, despite heat waves, droughts, and rising sea level. The world would, after all, be less crowded by 2 billion people.

The 1980s? Less crowded by 3 billion people. Technology would be familiar, although some things would of course be missing. We can say much the same for the 70s and 60s.

The 1950s? I was a kid then, and I still remember the joy of getting our first TV, a Zenith black-and-white with a screen no bigger than that on my current laptop. But we had running water, central heating, indoor toilets, cars, and so on. With about a quarter of today's population, suburbs had more open space. Life at that level would still be familiar. If society devolved with the technology, back to Jim Crow, segregation, separate facilities for blacks and whites, and worse, many of today's white supremacists would be quite happy. The rest of us? Not so much.

The 1920s would still feel fairly familiar though we would surely consider it a mite primitive. We'd have to get kicked all the way back to 1800 to have it really feel like a collapse of civilization. Dropping from 7.6 billion people to 1 billion people would be drastic enough to deserve the label. Yet people managed then, and we could do it again. There are people who know how to raise their own food, build houses, cut wood for fuel, handle horses. Remember that country folk may be hit less hard than city folk.

And of course, as soon as the needs of survival were handled, we'd be working hard at bringing back the electric grid, the Internet, and all the rest that went away when we couldn't maintain it.

Our needs would be less, with such a drastically reduced population. We wouldn't have to bring it all back right away. We wouldn't even have to bring things back in the same form. Renewable energy, for instance, would handle the needs of the smaller population just fine.

With luck, we might also face the future more intelligently than we have so far. We might, for instance, heed the lesson of the collapse, that the world has a carrying capacity, and let go of the mania for growth in population and development that got us into this mess. Instead, we could apply our resources to maximizing the well-being of the people, all of them, everywhere.

We could, in other words, take advantage of the collapse to start over and get it right this time around.

You are of course free to consider me far too optimistic.

ENERGY

Global warming is real. It's happening now, and it's going to get a lot worse. If we ever decide to do something serious about it, we will have to stop using fossil fuels. That means using something besides gasoline to run our cars—such as electricity or hydrogen. It also means using something besides coal for generating the electricity—such as nuclear, solar, geothermal, and wind.

Gimme Juice

Before the automobile there was the horse, and Lo! the streets were covered with horse puckeys and the air was filled with stink and flies. Most especially in cities.

People were happy to see the steam engine and the trains it enabled. But trains were restricted to tracks, emitted smoke and sometimes sparks, and had to haul large amounts of firewood or coal for fuel. Road-worthy steamers showed up too but never really caught on; the famous Stanley Steamer of the early 20th century was a Johnny-come-lately.

What really got rid of the horse puckeys was the discovery of petroleum. It and its derivatives such as gasoline, diesel oil, fuel oil, and kerosene are energy-dense, ash-free, and easy to transport in tanks and pipes, as well as abundant and relatively inexpensive. They are thus well suited to powering internal combustion engines, and it did not take long for inventors to come up with affordable automobiles to take advantage of them.

Coal, petroleum (and its derivatives), and natural gas are fossil fuels. Eventually we realized that for all that they get rid of horse puckeys in the streets, they too have problems such as air pollution, landscape destruction, and oil spills, as well as urban sprawl and strip malls. They also emit huge quantities of carbon dioxide, a major contributor to global climate change, AKA global warming, which threatens to raise sea levels, drown coasts, hammer agriculture with droughts, encourage the spread of disease from the tropics into the temperate zones, and so on. The litany of ill effects is long enough to make one wonder whether horse puckeys might not be preferable.

Fortunately, there's an alternative. Electric cars came on the market at roughly the same time as the early automobiles. They relied on heavy lead-acid batteries and had just enough range for Grandma to toodle around town.[1] And the batteries had to be charged with electricity generated by burning fossil fuels.

Today electricity can be carbon-free. It can even be cheaper when generated without fossil fuels; in December 2018, a massive Chinese solar power plant was pricing its output at less than the cost of coal-generated electricity.[2] In the U.S., renewable energy, mostly solar and wind, is now cheaper than electricity from 75% of coal plants.[3] By 2030, renewables will outcompete fossil fuels almost everywhere: "renewables are now the cheapest form of new electricity generation across two thirds of the world—cheaper than both new coal and new natural gas power."[4]

Not that solar and wind are the only carbon-free energy sources. There are also geothermal, hydroelectric, and tidal power sources. And in 2011, the World Wildlife Federation noted in a report that energy derived from renewable, carbon-free sources "has the potential to meet the world's electricity needs many times over, even allowing for" the way the sun only shines for part of the day and the wind doesn't blow all the time. Supply can be evened out with a wide-scale "smart grid." Many other reports concur. Given the political will, we could replace 80 percent of fossil fuel use by 2050 and go up from there, taking perhaps a century to reach 100%. It's doable.[5] Given the problems we have come to see in our dependence on fossil fuels, it's necessary too.

We do not yet have a unified global effort to stop using fossil fuels. If it were up to Big Energy, we might never see such an effort, for as long as they can continue to sell coal, oil, and natural gas and the electricity, gasoline, and home heating oil derived from them, the longer they can continue to make bazillions of dollars. It's not hard to understand why they buried their own research showing that climate change was coming and then tried to cast doubt on the work of other researchers.[6] They've also done their best to buy the politicians who might have bent public policy in a wiser direction.[7]

That may be changing, however:

—In 2015, the G7 countries called for an end to fossil fuel use by the end of the century.[8]
—In 2017, the UN called for the G20 nations to phase out fossil fuel subsidies by 2020.[9]
—In 2018, California passed a law requiring newly built homes to have solar panels.[10]
—Also in 2018: California announced plans to phase out fossil fuels by 2045.[11]
—In 2019, Germany announced that it would phase out coal use by 2038.[12]

It's not unified, and it's not global. But it does look like some momentum is gathering.

Meanwhile, electric cars are gaining momentum too. In California, one in ten of all new cars sold is an electric, and the state has half a million of them in total. Twice that many have been sold in the United States as a whole.[13]

They have a long way to go to take over, and they may never totally eliminate gas-burners. Cars can stay on the road for decades, especially if they have a fanbase willing to spend time and money on renovations. People still drive Model T's, after all.

But hybrids are getting more popular. They have a gas engine that either keeps the battery charged or kicks in when the battery gets low.

So are all-electrics, and though they have come a long way from the old Grandma-toodlers, they still are weak on range. It helps when employers and parking garages—and even municipal parking lots—set up charging stations. But that's not as convenient as being able to gas up at a filling station.

To get that kind of convenience we need better batteries, faster charging, and "filling" stations equipped to charge those batteries in no more time than it now takes to fill a tank. Another approach is a battery swapping system—when you pull into the station, they pull out your exhausted battery and then slide in a fully charged one. To make it work requires standardized battery packs that fit all cars (just as gasoline does). Making the batteries lease-only (like the old Bell phones) would keep people from gaming the system by swapping a decrepit old battery for a new one (and the batteries *are* expensive). This approach also requires redesign and re-equipping of "filling" stations—and there are a lot of them.

Another approach relies on hydrogen.[14] If we have enough solar, wind, and other electrical generation capacity to have surplus at certain times of day (demand is low at night, for instance), the surplus electricity can be used to electrolyze water (split its molecules) into oxygen and hydrogen. Hydrogen can be burned directly in an internal combustion engine, so it can be (and has been) used as automotive fuel. It can also be processed through a fuel cell to generate electricity. Since it's a gas, it can be stored in tanks and moved around through pipes, so it might be easier to adapt filling stations to using it. People complain that hydrogen is flammable—but then so is gasoline, with the added problem that a gasoline leak stays on the ground, around your ankles, while a lighter-than-air hydrogen leak rises quickly into the sky.

It took about half a century to build the automobile infrastructure up to a network of filling stations and paved roads that we would recognize today. We didn't build the Interstate system till the 1960s! With electric or hydrogen cars we just have to retool the filling stations, as well as build more renewable electricity sources. Much easier, especially since the timeline people are talking about is also a matter of decades.

Will we do it? Ask your local politician, if you can find one without a pocketful of Big Oil's money.

Actually, why isn't Big Oil doing it? It would be expensive, sure, but they could have a lock on selling what makes your car go. At least until the independents get in on the action.

Perhaps they are waiting for demand to go up. In 1951, carmakers sold over five million gas-burners. Going from one million electrics sold so far to five million a year might do it. But starting now to retool the filling stations and build more wind and solar farms would surely get us there faster.

Let It Glow

Once upon a time, nuclear power was pitched as promising electricity too cheap to meter. It didn't work out that way, for the power plants turned out to be absurdly expensive.

There were several reasons for this. One was the perception that nuclear power was dangerous, largely (at first) because after World War II "nuclear" meant "bombs." The Cold War didn't help, for we all then lived in fear of nuclear Armageddon. As a result, every nuclear power plant had a new set of safety measures, making each plant a one-off engineering challenge. Cost-saving standardization of design played little part.

Another reason for expense was years-long construction delays caused by redesigns and by public protests against putting the plants anywhere near anyone's back yard. This put off being able to start paying back loans, raised interest costs, and scared off investors.

Safety measures worked well when needed, at least until 2011. Three Mile Island, often cited as a horrific example of what can go wrong, released very little radioactive material to the environment. The Chernobyl disaster occurred when safety measures were ignored. In both cases, human error was more to blame than the technology itself. In 2011, however, a tsunami triggered by an undersea earthquake overwhelmed safety measures, destroyed several reactors, and spread contamination over large parts of Japan.[1]

And then there's the nuclear waste problem, whose icon is every reactor's "swimming pool"[2] full of spent fuel rods and glowing blue with Cerenkov radiation. Where do you put the U.S.'s 100,000 tons of intensely radioactive, toxic material (military and civilian, as of 2015) where it has no chance of getting into the environment and making people sick? The question became especially critical after a few notable slip-ups (including a Soviet waste dump that exploded and contaminated 20,000 square miles of land[3]). The answer approved by the U.S. government was underground storage, but that has proved so politically controversial that the only approved site, at Yucca Mountain in Nevada, has been closed. Reactor waste now sits in giant storage casks, mostly on site.[4]

It is no surprise that many people argue vehemently that nuclear power has no place in the modern world. It is just too dangerous.[5] As a result many countries are shuttering their nuclear power plants. Yet at the same time, there are calls for a nuclear revival or renaissance or resurgence. Advocates of nuclear power argue that the problems with nuclear power define their own solutions—standardized designs and expedited permitting are often mentioned. So are innovative reactor designs. "Fail-safe" reactors are designed to shut down without destroying themselves and releasing radioactive material. SMRs or small modular reactors[6] are suited to places that don't need huge, expensive reactors, such as developing countries,[7] as well as to applications with relatively modest power requirements, such as desalination. Among their advantages are size, price, and standardization—they can be built in factories, in quantity, and shipped to sites anywhere in the world. They have even attracted the attention of entrepreneurs.[8]

There are also benefits to be had. Nuclear advocates stress that nuclear power has zero carbon emissions (unlike coal and natural gas-based power), which can help us reduce the human contribution to global warming and—if we build the necessary thousands of reactors—save the world. If we choose to rely for the bulk of our energy needs on renewable sources (such as wind and solar), nuclear can provide essential "baseload" power for when renewables aren't up to the job, as at night and on windless days.

Richard Rhodes, "Why Nuclear Power Must Be Part of the Energy Solution," *Yale Environment 360*, July 19, 2018,[9] argues that nuclear is actually safer than most energy sources and is needed if the world hopes to radically decrease its carbon emissions. Laura S. H. Holgate and Sagatom Saha, "America Must Lead on Nuclear Energy to Maintain National Security," *Washington Quarterly*, Summer 2018, argue that economic, environmental, and technological factors are converging to make commercial nuclear power a sensible electricity source for countries looking to decarbonize their economies without sacrificing growth. To ensure stronger nuclear nonproliferation and security safeguards, the United States must once more embrace the commercial nuclear market.[10]

But there is still the nuclear waste problem. Where *do* you put thousands of tons of intensely radioactive, toxic material where it has no chance of getting into the environment and making people sick?

Deep borehole disposal is an idea that dates back to the 1950s. In its latest version,[11] it means drilling a hole five kilometers deep, installing a liner, dropping canisters of waste into the bottom two kilometers, and sealing everything up with concrete and clay plugs. You need to drill in solid rock without many cracks and without anything like oil or gas that someone in the future might wish to drill for. Fortunately, there is a lot of that kind of rock all over the world. And you need a lot—the U.S. alone would need about a thousand boreholes, costing $40 million each.

You go that deep, and there is very little chance that the waste will come back to haunt you. It seems to be the safest waste-disposal solution anyone has yet come up with. And, unlike back in the 1950s, we have the drilling technology to make it work. Nevertheless, the public is leery of this technique. Government efforts to find a site for a demonstration project—drill, drop non-waste down the hole, and seal it up—had to be cancelled in 2017 because the Department of Energy "had trouble finding a location for a borehole demonstration project. Communities feared that the test drilling site might later be used for nuclear waste disposal."[12] The DOE is still looking for suitable demo sites.

There are two genuine drawbacks. Once you've dropped the nuclear waste down the borehole, it would be extraordinarily difficult to get it back if you decided that you needed to extract components for crucial uses, and current U.S. law requires that the waste be retrievable. Here the solution is obvious—change the law. The second drawback—and it applies to any method of nuclear waste disposal—is perceptual. The public's first thought, on hearing that someone wants to put nuclear waste anywhere nearby, is "OHMIGAWD!!! WE'RE ALL GONNA DIE!!" Or, with apologies to Tom Lehrer and the Cold War, "We will all glow together when we go."

Coal, oil, and natural gas are more likely to kill you, thanks to global warming. We dump fossil-fuel waste into the air, where it traps solar heat, increases the strength and frequency of big storms, droughts, and heat waves, and melts polar ice caps. When it gets bad enough, perhaps people will yell "OHMIGAWD!!!" about fossil fuels.

Will they learn? Will they decide nuclear's not so bad after all? Its problems can be solved. The waste can be put where it's out of reach, and where we're out of range of its nasty effects.

We just have to look at the real risks. Take our fear and let it go. Dump the waste down a deep, deep hole, and let it glow.

Shine On!

We have known how to use solar energy for a long, long time. It's as simple as facing a house to the south to catch the winter rays. If you want hot water, put a black-painted 50-gallon drum on the roof. Or take a long black hose and snake it back and forth on the roof, as I once saw in Maine.[1] If you want something fancier for heating household water or even living spaces, you can install special panels.[2] If you need more intense heat, try a Fresnel lens.[3]

If you want to use the sun to generate electricity, there are two basic methods. One that has been used by utilities relies on mirrors to concentrate sunlight on a boiler, generate steam, and spin the turbines of electrical generators. The largest and most recent example is the Ivanpah solar farm in California's Mojave Desert; it was designed to produce over a million megawatt-hours of electricity per year."[4]

You can also use solar cells, which lend themselves to much smaller applications. They can be deployed anywhere a flat surface faces the sun—on flashlights, backpacks, rooves, cartops, free-standing frameworks, and even roads.[5] The Tesla company is marketing solar roofing tiles.[6] And in 2018, California passed a law requiring newly built homes to have solar panels.[7]

It's not hard to use them at home. Businesses such as Sungevity, Sunrun, and Vivint will install solar cells on home rooftops, let customers use some of the electricity generated, and sell the rest to utilities. People can also buy solar systems outright. And once you pay off the up-front costs, you have free electricity for the rest of your days. For nights, you need battery backup.

Costs of solar panels have been falling; solar electricity may even be competitive with fossil fuel-based electricity before long. Demand has been rising. And this is giving utilities fits, for the law requires that they maintain an appropriate infrastructure (power plants, poles, transformers, wires, repair trucks) to meet demand, and if demand is falling, they lack the necessary revenue. Peter Kind, executive director of Energy Infrastructure Advocates, argues that increased interest in "distributed energy resources" such as home solar power and energy efficiency, among other factors, is threatening to reduce revenue and increase costs for electrical utilities. In order to protect investors and capital availability, electrical utilities must consider new charges for customers who reduce their electricity usage, decreased payments to homeowners using net metering (selling surplus solar electricity back to the utility), and even new charges to off-grid users of "distributed energy resources" to offset "stranded costs" (such as no longer needed power plants).[8] That is, if the utilities lose customers, they must either charge the remaining customers more or find a way to bill non-customers. In Florida, local government seems prepared to help by preventing people from disconnecting from the grid.[9] In Hawaii, utilities restrict or block rooftop solar. In other states, they are pushing hard to do the same.

If the utilities cannot protect their business model, they say, they face a "death spiral" of falling revenues and rising liabilities.[10] They are doomed.

Are they really? A major drawback to home solar is that the sun doesn't shine all the time. You can make up for that to some degree with an electrical grid (the network of power plants, power lines, substations, and computerized control stations that controls the distribution of electricity even when disrupted by storms, power plant failures, and other problems) that stretches across enough time zones to bring electricity in from someplace where the sun is shining or the clouds are gone.[11] If you can't span enough time zones or you have a more local grid, you need traditional gas-fired, nuclear, and hydroelectric power plants to fill in the low spots in supply, on demand.[12] So there is still a need for the traditional utility. The question is whether the need is great enough to justify the costs. It may make more sense for homeowners to buy battery backup systems. They don't *have* to get used to the sort of electricity supply that was more the norm in the 1930s.

A more important question may be whether any business has a right to expect to keep making money and attracting investors the way it always has. The oil industry has been fighting this battle for years, and certainly oil wells, refineries, pipelines, tankers, and gas stations represent "stranded costs" if we ever swear off the stuff and go all electric.

Technological change happens, and it disrupts existing business models. It seems rather nervy, not to mention desperate, for a business, when this happens, to try to charge noncustomers for a service it is not providing them, on the grounds that they might need it someday and the business has to stand ready to meet the possible demand. Fortunately, the oil industry isn't proposing to charge electric car owners for the gasoline they don't use, though if the electric utilities can make their grab stick, it might reconsider. So might the tobacco companies, donut shops, and even gyms.

Surely there are other ways to stay in business. Rather than digging in the corporate heels, how about adapting to change? Perhaps the electric utilities could sell complete home solar packages instead, with both battery and grid backup. Or cover shut-down coal fields with solar panels to generate electricity more cleanly. Or invest in wind, geothermal, tidal, and other renewable energy sources.

These are, after all, changes we will have to make anyway, and the sooner the better. Global warming is a serious threat, and it will force us to stop using fossil fuels. Eventually. I suspect it won't happen till the consequences are much more apparent.

So far, the utilities are choosing to mount a "war on solar" instead.[13] Fortunately, and despite the Trump Administration's tariffs on imported solar cells, solar is winning. At least so far.[14]

Energy

POLITICS

Some of the issues that affect our future are more political than technical. They have technical components, to be sure, but the primary arguments are about policy, as in the cases of restricting potentially dangerous research, preparing for disaster, or seeking a way to avoid actually doing anything about a problem. And then there's the business of trying to shout down criticism by calling the media "enemies of the people." If that continues, we could be in as much trouble as any more technical issue could produce.

Be Prepared

Most of these little essays deal with matters of science and technology. My aim has not been to add to the flood of techno-hype. I have mostly chosen to point out potential problems and—when possible—potential solutions. Perhaps unfortunately, the realm of solutions is not the realm of science and technology, but of politics. And politics is more complicated than science and technology.

At one point, I mentioned that some of the problems we face—global warming, overpopulation, and future food supply, for instance—are predictable. We can have some idea of when we will need solutions, and we can have the solutions ready. Or not (thanks to politics).

But some problems are less predictable. We have no idea when the next big earthquake or volcanic eruption or pandemic will happen. We do, however, know that earthquakes, volcanic eruptions, and pandemics, as well as hurricanes and tsunamis, happen fairly often and can hammer large numbers of people.[1] We have therefore invested in weather satellites and computer models that let us watch developing storms and predict their tracks. Early warning systems for earthquakes, volcanic eruptions, and tsunamis are less well developed and implemented. Indonesia's tsunami warning system failed in December 2018, apparently because it was designed to listen for earthquakes (which often trigger tsunamis) rather than volcanic activity.[2]

These disasters can't be prevented. Money is spent on warning and monitoring systems because of their frequency and the damage they cause, both to lives and to property. Also damage estimates can be based on past experience and compared with the costs of warning and monitoring systems. The cost-benefit analysis beloved of economists is both doable and favorable.

Cost-benefit analysis breaks down however, when the disasters are rare and large, chiefly because though the costs of early warning systems and even prevention programs are knowable, the benefits depend on what is known as the discount rate. To see how that works, consider a disaster a century from now that will do $1 billion worth of damage. If you invest $1 million today at 7 percent interest, the money will double every ten years and in a century you will have the billion you need to fix the future damage. Therefore you cannot justify spending more than $1 million today to prevent that future disaster. The interest you earn on that money is the discount rate. If the rate is worse or better than 7%, the money you can justify spending changes accordingly.

It is at best awkward to try to put a dollar value on lives lost in a disaster, but most attempts to do so use numbers between one and ten million dollars per life.[3] A billion-dollar disaster can thus mean as few as 100 deaths, with no property damage. If the disaster involves millions of deaths, as well as extensive property damage, the bill approaches infinity and cost-benefit math just doesn't work.

Are there any possible disasters *that* bad?

Consider the Yellowstone Caldera or supervolcano.[4] It last blew some 600,000 years ago, and volcanologists assure us that it isn't about to do it again any time soon. But if it did, most of North America would see ash fall, and the global climate would be chilled, affecting food production.

It's a maybe-someday sort of disaster. The cost would be huge, you can't do anything to prevent it, and maybe suitable monitoring (future monitoring systems will surely be better than today's) could see it coming. Given enough warning you could evacuate nearby populations (several states' worth) and stockpile food. But how much can you justify spending today? And on what?

It would seem reasonable to spend money on better highways in the region to aid evacuation. They would also be useful in case of lesser eruptions in the region, and they would aid travel and commerce generally. A similar multi-benefit approach appears if you spend money on improved monitoring systems, for they can be valuable anywhere in the world where there is a volcano. Unfortunately, global benefits often don't factor into national political decisions.

Is there anything worse than a supervolcano?

Of course there is! Sixty five million years ago, a comet or asteroid 10 kilometers (6 miles) in diameter struck near what is now Chicxulub, Mexico: The results included the extinction of the dinosaurs (as well as a great many other species). Ancient history, you say? Meteor Crater in Arizona, almost a mile across, was created some 50,000 years ago by a meteorite just 150 feet in diameter.[5] An even larger crater was recently discovered underneath the ice that covers most of Greenland; the date isn't certain, but it may have happened only 13,000 years ago and caused the thousand-year-long global chill known as the Younger Dryas.[6]

Such events are rare. Chicxulub-scale events are very rare; a hundred million years may pass between them. Meteor Crater-scale events may occur every thousand years, releasing as much energy as a 100-megaton nuclear bomb and destroying an area the size of a city. A Greenland-sized event may happen every few million years; if one hit North America or Europe, it could destroy entire nations. If it hit in the middle of the Atlantic, the resulting tsunami would wash away entire seaboards.

More modest impacts also count as major. We lucked out in 1908 when a rock a third the size of a football field exploded over Tunguska, Siberia.[7] A similar event today—especially if it happened over a major population center—would be one of the greatest catastrophes in recorded history. And according to NASA Administrator Jim Bridenstine, in his keynote talk at the International Academy of Astronautics Planetary Defense Conference, April 29, 2019, you can expect a major asteroid strike—Tunguska-scale or bigger—during your lifetime.[8]

How do you do a cost-benefit analysis on preventing disasters like these? It is not really possible. Whatever the prevention cost, the benefit is "everything." Yet money is finite. You really do need to consider multi-benefit measures that can be justified in more than one way.

Warding off meteorite impacts is conceptually simple:

1. Spot them on the way. This means using telescopes to inventory space rocks than come anywhere near Earth. According to NASA, there are some 4,700 asteroids more than 100 meters (330 feet) across, of which only 20-30 percent have actually been discovered so far. Greg Easterbrook, "The Sky Is Falling," *Atlantic,* June 2008, argues that human society faces so much risk

from asteroid and comet impacts that Congress should place a much higher priority on detecting potential impactors and devising ways to stop them.

2. Develop ways of deflecting them so they miss Earth. Many methods are discussed in the "National Near-Earth Preparedness Strategy and Action Plan" prepared by the U.S. Interagency Working Group for Detecting and Mitigating the Impact of Earth-Bound Near-Earth Objects (NEOs) of the National Science and Technology Council, June 2018.[9]

Most deflection methods require a much more ambitious space program than anything the U.S or other countries have today. It would be immensely expensive, and no politician would be willing to try justifying it with a cost-benefit analysis. It would cost trillions, and in the worst case the damage estimate is very simple: Everything. Discount rate is meaningless.

Clark R. Chapman, in "What Will Happen When the Next Asteroid Strikes?" *Astronomy*, May 2011, argues that though the consequences of an impact would be catastrophic, efforts to prevent the impact would be futile. Better, he says, to plan for evacuating target zones and providing shelter and food for the refugees.

We don't have to buy that. A more ambitious space program would provide a great many benefits while we wait for an impact to threaten us. We could mine asteroids, build solar satellites to beam energy down to Earth, set up research bases on the Moon and Mars, and perhaps even—eventually—build colonies on other worlds.

Of course, supervolcanoes and cosmic impacts don't pose threats any time soon. It's all long-term stuff—though you really should keep an eye on NASA's Coming Attractions list.[10] And politicians don't think long-term.

Supervolcanoes and cosmic impacts do, however, become short-term concerns once we see them coming. With the first, we just need time enough to get out of the neighborhood. With the second... well, how much advance warning do we need to build up, in a great rush, a space program capable of warding off a big rock? Five years is surely not enough. Ten? Twenty-five? Fifty? Surely, we need all the warning we can get. With no warning, disaster could come at any time. With not enough warning—that's not much better. It doesn't matter how much it costs. It matters only that we can do it—in time.

I mentioned pandemics above. They are catastrophes. They are smaller catastrophes than supervolcanoes or space rocks, and they are unpredictable. But like the larger catastrophes, we know they will happen. The question is not whether but when.

The worst pandemic ever to have affected the Earth's human population was probably the Black Death (bubonic plague), which when it first reached Europe in 1347 killed a third of Europe's population.[11] In later attacks, it killed many more.

We have also had deadly flu pandemics. The 1918 one gets most of the attention, but we have also had smaller ones in more recent years.[12]

The most recent pandemic was the 2020 Covid-19 pandemic, whose full toll has not yet been tallied. It began in China as early as November 2019 and promptly began sweeping across the globe, causing over 6 million infections and almost 400,000 deaths by the June 1, 2020. The cost of those deaths, at $10 million each, is 4 trillion dollars; national lockdowns cost trillions more in lost economic activity.[13] The final totals are likely to be much higher.

Variously known as SARS-CoV-2, Covid-19, or coronavirus, the virus belongs to the same family as cold viruses, SARS (Severe Acute Respiratory Syndrome), and MERS (Middle East Respiratory Syndrome). Like the last two, it is a zoonosis, meaning a disease that has leaped from its normal animal host—probably bats, in this case--to humans. Such leaps are responsible for many of the most devastating diseases in human history, including smallpox, measles, bubonic plague, and more.

And as the human population grows, it expands into new territory and comes into contact with more wildlife, and with new disease organisms.

In other words, it may not have been possible to predict the appearance of the Covid-19 pandemic specifically, but it was clearly possible to predict the appearance of new, potentially serious zoonoses and pandemics. This is why the United States' Centers for Disease Control (CDC) has maintained early-warning stations around the world to watch for new diseases, as well as pandemic response teams in both the White House and the CDC.

Spotting new bacterial and viral diseases early is crucial. As we learned with the 2020 Covid-19 pandemic, an essential element of coping with the crisis is being able to test for the disease. Since it takes time to develop tests, the more lead time the better. Unfortunately, you can't start developing tests until you know what to test for. You can predict that pandemics will happen; you cannot predict what they will be. Therefore you need to spot them as soon as they show up, *before* they become widespread problems.

The Trump Administration dismantled the early-warning stations and pandemic response teams—because they were supposedly a waste of money--in a perfect illustration of how NOT to be prepared for disaster.

As the pandemic has developed, it has also become clear that the world's medical facilities are not adequate to the task. There are not enough hospital beds, not enough protective masks and gowns, not enough respiratory ventilators, and not even enough medical staff to make up for those who fall ill. The National Stockpile of medical supplies was completely inadequate. This tells us that in order to be prepared for such a disaster, there must be what in normal times would be considered excess capacity. We should note that providing "excess" capacity is very much a multi-benefit approach. Preparedness for pandemics is also preparedness for other kinds of disaster. It would also mean that in ordinary times health-care facilities would be more available to everyone.

Unfortunately, our profit-driven health-care system sees excess capacity as an unnecessary burden on profits. Perhaps it would be wise to return to that time (before 1973 when President Nixon gave the health-care industry permission to make profits) when health care was viewed as a public service. Given the readiness of retired medical people to leap back into the fray during the 2020 pandemic, that ideal is not dead.

Emergency preparedness costs money. It's expensive. But it is surely more important than profits.

If It Quacks like a DURC...

That title may sound silly, but it has a point. DURC stands for "Dual-Use Research of Concern," and it refers to research whose dissemination may permit people to do harm. Some say that dissemination should therefore be restricted in some fashion short of declaring it "Classified" or "Top Secret." Saying "If it quacks like a Top Secret" might be more apt, but it wouldn't be nearly so much fun.

As a term, DURC came to prominence in connection with some alarming research on bird flu. As its name implies, "bird flu" affects primarily birds.[1] It can also affect mammals, including humans, but it does not spread as easily among mammals as it does among birds. In particular, unlike the ordinary flu that we get vaccinated against every year, it does not spread via droplets sneezed or coughed into the air. Thus even though it may infect a few hundred people in an outbreak, and kill a sizable proportion of its victims, it holds little potential to cause a pandemic (widespread epidemic) akin to the 1918 flu, which killed at least 50 million people.[2]

A few years ago, two research teams led by Ron Fouchier of Erasmus MC in Rotterdam, the Netherlands, and Yoshihiro Kawaoka of the University of Wisconsin, Madison, modified the H5N1 bird flu virus so that it could move more easily through the air between ferrets (which respond to flu much as do humans). This made it potentially more hazardous to humans. When in 2011 the two groups submitted papers to *Nature* and *Science* explaining how they had modified the virus, the journals accepted the papers but refused to publish them until the United States' National Science Advisory Board for Biosecurity (NSABB) could decide whether key details should be removed or "redacted" from the papers (and made available only to researchers with a clear need to know) in order to prevent terrorists from learning how to create a flu pandemic.[3] Alarmed critics bloviated that the modified virus was an "Armageddon virus."

The key details could be essential to identifying a more infectious version of the natural virus before it caused a pandemic,[4] but nevertheless the NSABB recommended redaction. It soon reconsidered, though, and in 2012, the two papers—unredacted—were finally published.[5] Later that year, the National Institutes of Health (NIH) proposed stringent reviews of whether similar research should receive government funding, or even be classified.[6] Research resumed in 2013, but funding was put on hold from 2014 to 2019.[7]

The DURC question remains live,[8] even though researchers seem to have kept their heads down in the last few years. At least, the dates on the examples below are a few years old.

But really, is it true that some research is too dangerous to permit or, if permitted, too dangerous to publish? An essential component of the scientific method is publication. Form a hypothesis, do the necessary experiments to determine whether the hypothesis is wrong, and then tell the world so others can check your work. If you work for a corporation, however, your reports might never be seen by anyone outside the company. If you work for the military or on defense contracts, your reports might be classified and, again, never seen by anyone without the appropriate security clearance.[9] National security can come into play in other ways as well, as it did forty years ago when mathematicians working on cryptography ran into efforts at control by the

National Security Agency.[10] Getting past efforts at control is what spies exist to do.

Aside from proprietary, military, and national security contexts, is there any real reason why any knowledge, new or old, should be kept from the public eye? If you consider what one can do with very easily available knowledge, you might wonder. High school chemistry is enough to figure out how to build a bomb out of things found under the kitchen sink.[11] High school biology can reveal how to make a whole prom ill by putting sewage in the punch or how to breed (by natural selection) antibiotic-resistant bacteria. A college microbiology text... A mycology (fungi, including deadly mushrooms) text... Well. And these days it's all on the Internet, too.

A paper by Lawrence M. Wein and Yifan Liu, "Analyzing a Bioterror Attack on the Food Supply: The Case of Botulinum Toxin in Milk," dealt with how terrorists might attack the U.S.'s milk supply (and therefore with how to safeguard it). It was scheduled for the May 30, 2005, issue of the *Proceedings of the National Academy of Sciences* until the Department of Health and Human Services asked the NAS not to publish the paper on the grounds that it provided "a road map for terrorists and publication is not in the interests of the United States." The journal eventually published the paper anyway.

The rapid advances in genetics and DNA manipulation have provided even more examples. The July 2002 report that researchers had successfully assembled a polio virus from biochemicals and the virus's gene map roused fears that if you can do it with polio, you can do it with smallpox. It also led directly to the formation of the NSABB. The October 2005 report that researchers had reassembled the deadly 1918 flu from synthesized subunits[12] soon led to calls for researchers to censor their own work.[13]

It is easy to say that all knowledge should be freely available. It is, after all, not what one knows, but what one does with what one knows, that creates problems. Unfortunately, the things that quack like a DURC tend to be the obvious things. Most of the problems that result from scientific and technological knowledge are ones that we do *not* see coming. The Green Revolution[14] of the 1960s solved a looming hunger problem, but it permitted population to triple (so far) and thus recreate the original problem on a larger scale. The invention of the automobile gave us traffic jams, road rage, and urban sprawl. The invention of the transistor in 1947 led to computers, smartphones, video games, the Internet, and Artificial Intelligence, with issues of privacy, obsession, fake news, and unemployment, among others. The discovery of how to splice genes in the 1970s led to genetic engineering, GMO crops, manipulation of human embryos, and eventually transhumanism (all of which some insist are problems). And so on.

Most of us would say the benefits of scientific and technological progress outweigh the drawbacks. It is better to know more, and to do more, and to keep striving to know and do even more. We will stumble from time to time, but we will recover and move on.

But what about the bad actors, you say? The ones that DURC fans worry about? The terrorists and the deranged? Synthesizing a smallpox virus, or devising something even worse, requires advanced knowledge (even though some of the necessary equipment, such as DNA sequencers, is available on eBay). A foreign government is more likely to do it, and they have the personnel to work out the methods on their own (that it *can* be done is obvious; you just need the gene map, which is what DNA sequencing machines give you).

Making mischief from basic textbooks is both more likely and unstoppable. Those textbooks are *everywhere,* including your grandmother's attic.

Rather than restricting research and publication it might be better to create an agency whose mission is to imagine all the awful things that could result, deliberately or not, from available knowledge—textbooks, Internet, research papers—and then to figure out how to cope, adapt, or fix the problems.

Dealing Green

The 2018 midterm elections brought to the U.S. House of Representatives a flood of new Democratic talent with a propensity to say "We need to do things differently."

This propensity has taken notable form with House Resolution 109,[1] better known as the "Green New Deal," introduced in the House by Alexandria Ocasio-Cortez (D-NY) on February 7, 2019. The Resolution has been attacked as unaffordable, impractical, a threat to the economy, and of course (gasp!) socialist.[2] The Green Party loves it, saying that it "will convert the decaying fossil fuel economy into a new, green economy that is environmentally sustainable, economically secure and socially just."[3] The Sierra Club approves.[4] Some commentators are saying the Democrats finally have a worthy cause.[5]

What does the Resolution actually say?

It begins by noting that global climate change is real, human activity is the dominant cause, and it threatens immense damage to biological diversity, public infrastructure, the national economy, national security, and human lives. This much is not arguable. The scientific consensus is clear about the data and their implications, and even the Trump Administration's own reports (e.g., the November 2018 Fourth National Climate Assessment[6]) back it up. The Resolution then calls for the U.S. to take the lead in reducing net global carbon emissions to zero by 2050.

Whether reducing carbon emissions that much that soon can actually be done is not at all certain, but the Resolution is not a proposal for specific action or even legislation. It is a list of goals to work toward. Inability or failure to achieve a specific goal by a specific date is thus not a reason to dismiss the goal as unachievable. We can at least start working on it even if it may take longer than we plan or wish.

The Resolution next addresses social justice issues, noting declining life expectancy, wage stagnation, deindustrialization, anti-labor policies, and increased income inequality, with disproportionate impacts on "indigenous peoples, communities of color, migrant communities, deindustrialized communities, depopulated rural communities, the poor, low-income workers, women, the elderly, the unhoused, people with disabilities, and youth." There is thus a need for millions of good, high-wage jobs to improve prosperity and economic security for all Americans. There is also a need to counteract systemic injustices. Therefore the Federal government has a duty to reduce greenhouse gas emissions, create jobs, invest in infrastructure and industry, and ensure that future generations of Americans can enjoy clean air and water; climate and community resiliency; healthy food; access to nature; a sustainable environment; and justice and equity.

The Resolution then offers a long list of specific steps—including providing everyone with high-quality health care—to be taken over the next ten years to move toward these goals. Many of these steps have been recognized as necessary by environmentalists for more than half a century. Social justice issues have come to the fore more recently. All deserve attention in any society that believes a nation's government is supposed to benefit all its people, not just a small (one percent) subset of its people.

It will be expensive to do it all, but as the Resolution notes, we are facing a loss to the national economy of $500 billion a year by 2100, just from the effects of climate change, as well as a loss of $1 trillion in damage to infrastructure and coastal real estate. It may thus be more expensive not to do it all. Certainly, we should try.

Can it all be done? Much of it can, but at least one important component probably cannot, except in the very short term. This is the "providing jobs" component.

The jobs are needed in part because so much of American manufacturing has left the country for venues where labor is cheaper and regulations are fewer. Conservatives say we can get the jobs back by getting rid of regulations and warding off the push for a livable minimum wage. President Trump has done his best to get rid of regulations, but as soon as he's gone, they will be restored. They are there for good reasons—protecting worker safety, consumer health, clean air and water, and so on.

In the post-Depression era of the original New Deal, the nation provided jobs by creating the Civilian Conservation Corps[7] and the Works Progress Administration[8] to help people make it till the economy improved. But many of those manual-labor jobs—such as road construction—are now done with the aid of heavy machinery. If we put people to work on such jobs without the aid of the machinery, when they know the machinery exists, the feeling that the jobs are make-work jobs will be hard to escape. People need to feel that they are doing something useful.

And as I noted earlier in this book, employment is under threat by Artificial Intelligence (AI), meaning extreme automation. The impact of AI is already visible—even in fast-food restaurants—and projections are for it to become ever more pervasive. The need for jobs is going to increase tremendously. Very few people think it is possible to ban AI, though taxing it at a level that makes human labor more competitive might help.

Or is it really the need for jobs that will increase? Again as I noted earlier, it may be more realistic to recognize that the real need is for income, at a level that allows people to meet at least their basic needs, to buy goods, and to keep the economy alive. With neither jobs nor income, the economy crashes.

When the government provides income without jobs, this is a "Guaranteed Basic Income" program. It has to be paid for, of course, but taxes can be levied on those who have income above the basic level. A return to a more progressive income tax would also help. Nationalizing selected industries (healthcare, anyone?) might also be useful. So might deciding that the constitutional separation of religion and government should mean that religion does not deserve special tax breaks.[9]

Why didn't Alexandria Ocasio-Cortez include Guaranteed Basic Income in the Green New Deal? It's unavoidable, eventually. But at present it is a radical notion without roots such as other components of the Green New Deal have (the original New Deal, the original union movement, the Fair Labor Standards Act, the minority and women's rights efforts, and more). It also smells of anti-capitalist Marxist communism. Thus a national Guaranteed Basic Income program is at present very unlikely to make it past conservatives in Congress. After all, they say, if you don't have a job or are poor, it's your own fault, you're just lazy, or—that old-time Puritan ethic—you're a sinner and it's God's judgment on you, and it is not government's role to argue with God. It is also not government's role to argue with Profit. After all, the One Percent is responsible for a great many campaign contributions.

If the job-related impact of AI actually develops as projected, that will change, just to keep the economy working. Until then, the Green New Deal as it stands before Congress may be the best we can hope for as a list of worthwhile goals and a guide to future policy and legislation.

And let's face it: Even without a Green New Deal, protecting the environment, improving wages and working conditions, making sure housing and health-care are affordable, and trying to give everyone a fair break should be on the agenda.

Do It Now

On September 20, 2019, students and other groups in Australia, France, Germany, India, Tanzania, Uganda, Ukraine, the United Kingdom, the U.S., and many other nations around the world went on strike to protest global inaction on the climate crisis.[1] The protests continued on Monday, September 23, with activists blocking streets in Washington, D.C. The police were making arrests.[2]

The protests were timed to precede and accompany the U.N. Climate Action Summit 2019,[3] whose web page says:

"Climate change is the defining issue of our time and now is the defining moment to do something about it. There is still time to tackle climate change, but it will require an unprecedented effort from all sectors of society. To boost ambition and accelerate actions to implement the Paris Agreement on Climate Change, UN Secretary-General António Guterres is asking leaders, from government, business and civil society, to come to the 2019 Climate Action Summit on 23 September with plans to address the global climate emergency. The Summit will spark the transformation that is

*urgently needed and propel
action that will benefit everyone.
"Global emissions are
reaching record levels and show
no sign of peaking. The last four
years were the four hottest on
record, and winter temperatures
in the Arctic have risen by 3°C
since 1990. Sea levels are
rising, coral reefs are dying, and
we are starting to see the life-
threatening impact of climate
change on health, through air
pollution, heatwaves and risks
to food security."*

The closing press release[4] announced that:

*"77 countries committed to
cut greenhouse gas emissions to
net zero by 2050, while 70
countries announced they will
either boost their national action
plans by 2020 or have started
the process of doing so.
"Over 100 business leaders
delivered concrete actions to
align with the Paris Agreement
targets, and speed up the
transition from the grey to green
economy, including asset-
owners holding over $2 trillion
in assets and leading*

*companies with combined value
also over $2 trillion.
"Many countries and over
100 cities—including many of
the world's largest—announced
significant and concrete new
steps to combat the climate
crisis."*

Is this enough? The December 2019 United Nations Climate Summit (COP 25) declared a need for nations to be more ambitious about reducing carbon emissions and helping poorer nations cope with the impacts of global warming. A discussion of global carbon markets was put off till the 2020 Summit.

The issue is certainly urgent. New European climate models say a worst-case scenario may mean the global average temperature goes up as much as 7 degrees Celsius (12 degrees Fahrenheit) by 2100.[5] This means that the heat waves and storms that have plagued large parts of the world in recent years will become worse and more frequent. It means projections that large parts of the world—such as South Asia—will become unlivable for humans are more likely. It means that the ice of Greenland and Antarctica will melt faster, raising sea levels faster as a result.[6] We don't have long to get a grip on this.[7]

And we know what to do about it: Stop burning fossil fuels and emitting carbon dioxide. Shift the massive subsidies given to the coal, oil, and natural gas industries to wind, solar, and geothermal (which are cost-competitive in many areas). Suck carbon dioxide out of the atmosphere.[8] Plant trees. Encourage more cities, states, and countries, and even corporations to join the move to be carbon neutral before it's too late.

It won't be cheap. But failure will be far more expensive. New York City is already planning to spend $20 billion on sea walls and white rooftops to fight climate change.[9] And that's just one city. The global bill will be trillions. Just try to imagine what it would take to move a city like Miami to higher ground (of which Florida doesn't have much).

And failure looks more likely than success at the moment. In the U.S. the denialists still rule the roost—the Trump administration still claims global warming is a Chinese hoax intended to emasculate American industry (although Trump may be changing his tune, at least a little[10]).

The rest of the world has denialists too. When Brazil's president, Jair Bolsonaro, spoke to the UN's General Assembly, he reiterated his claims that the Amazon is not burning, despite data reported by his own government. It's all media lies, he said. He too is a climate denialist; he claims there is no need to protect the nation's forests to fight a bogus climate crisis. National sovereignty and economic development come first.[11]

But projections of future warming are consistently proving to be too conservative. In other words, if the experts are saying the future looks bad, the reality will be worse. We should have acted decades ago, when we first knew what was coming. But we didn't. No one wanted to disrupt business as usual—there was far too much money to be made from coal, oil, and natural gas, and politicians could count on generous campaign donations from those industries.

Now young people are recognizing the urgency of the issue and massing to protest inaction. Some old folks share the sense of urgency, but not enough. As Swedish teen climate warrior Greta Thunberg has said, the house is on fire and it is time to panic.[12]

At 16, Greta Thunberg is the example who shows that the kids protesting now don't have to get any older to make their voices heard. But they do have to get older—decades older—to move into decision-making positions in government and industry. And we may not have decades—more than one analyst sets the do-or-die date as 2030.[13]

Fortunately, the kids will be able to vote long before then.

All they'll need will be suitable candidates.

If those candidates do not appear, in due time today's activist youth will *be* those candidates. And then we will certainly see action.

Let us hope that that action will not be too late.

Who's the Enemy?

For many years, my daily commute took me through the town of Albion, Maine, the birthplace of Elijah Parish Lovejoy (1802-1837).[1]

Lovejoy comes to mind because of the Trump Administration's repeated charges that journalism ("the media") is the enemy of the American people. Certainly the people who murdered Lovejoy would have agreed.

You see, he was a newspaper publisher and journalist in Alton, Illinois. He was also a Presbyterian minister with a strong moral sense that led him to oppose slavery. In other words, he was an abolitionist surrounded by people who thought slavery was just fine and anyone who opposed it was an "enemy of the people."

Anti-abolitionists destroyed his printing press three times in St. Louis before he crossed the river to Alton in 1836 and started a new abolitionist newspaper. It was not long before the mob came for him again, and this time they shot him.[2]

Since then, his name has been memorialized by being applied to schools, buildings, and awards such as Colby College's Elijah Parish Lovejoy Award.[3] Colby, in Waterville, Maine, was his alma mater. It initiated the award in 1952 to honor his heritage of fearlessness and freedom and to link journalistic and academic freedom.

Was Lovejoy an enemy of the people? The people certainly thought so, for he criticized them and their values, and he did so continually, despite strenuous opposition. Yet it was not long before his own abolitionist values were validated by the result of the Civil War.

The slaves were freed, although the pro-slavery faction has not yet vanished from public life. Nor have those who view criticism as enmity deserving of attack and even death.

Lovejoy is therefore worth remembering, even this long after his death. Journalists have a duty to expose corruption and evil, to criticize those responsible, and to honor those who work for the good.

Journalism is much more the conscience of the people than its enemy.

And if you think your conscience is your enemy, you have a problem.

CONCLUSION

Grandpa Did It. Why Can't I?

One of the classic objections to environmental regulations is that they interfere with the rights of a landowner. A person or a corporation resents being told that they cannot dump sewage or industrial waste into a river or lake or send clouds of smoke into the air. "After all," they say, "we always used to be able to do that."

There are also situations where the complainer is not a landowner, but a land user. A prime example is the rancher who grazes cattle on public lands. He pays a fee much, much lower than if he were grazing on privately owned lands. And he complains bitterly if told the fee is going up, or if the number of cattle he is allowed to put on the land is reduced in the name of protecting the land and streams. "After all," he cries, "We've been doing this for generations! How dare you defy tradition!"

It's not just a matter of property rights or grazing rights. That's how it is usually presented, as by the supporters of Cliven Bundy, who seized the Malheur National Wildlife Refuge in Oregon a few years ago. Bundy had refused for years to pay the necessary fees for grazing his cattle on federal lands in Nevada, claiming that the Federal government lacked the authority to manage public lands. The seizure led to a prolonged standoff with law enforcement, one death, arrests, and trials. Eleven of Bundy's supporters pled guilty; his sons were acquitted.[1]

But the issue is broader than the usual terminology suggests. The public lands affected include rangelands, forests, wilderness areas, and wildlife refuges. The legislation that governs them often specifies that they should be managed for "multiple use," meaning use for recreation, conservation, and exploitation (grazing, mining, logging, etc.). According to the U.S. Department of

Justice,[2] "For historical reasons, the statutorily sanctioned timber and grazing uses of [multiple use] lands have resulted not only in the expectation by ranchers and the timber industry that these uses will continue unabated, but [also] similar expectations in communities whose livelihood depends on the persistence of these uses."

The broader issue is thus that of "traditional expectations of land use," not property or grazing or dumping rights. "We've always done it this way. You have no business changing the rules!" Or, "Grandpa did it this way. Why can't I?"

Unfortunately, Grandpa[3] did a lot of things we don't do anymore. He rode a horse instead of driving a car. He raised his own food and cut his own firewood. He went hunting and fishing without worrying about bag limits. Depending on where he lived, he might even have attended lynching parties and killed annoying journalists.

Times change. Often, they change for very good reasons. In the case of environmental regulations, those reasons include that we have realized that the "traditional" ways had some thoroughly nasty side-effects, such as destroying wildlife and turning waterways into open sewers. In the 1960s, Cleveland's Cuyahoga River infamously caught fire. In 1952, smog killed 4000 Londoners. Add in oil spills, acid rain, ozone holes, and global climate change. We have good and sufficient reasons for telling people that they can no longer do things the way Grandpa did.

Note that it isn't just that Grandpa didn't know better. There were fewer people around in his day. If he was a rancher, he didn't graze as many cattle on public land. If he was an industrialist, he didn't generate as much crap to dump in the river. When America was expanding westward, the focus was on building industries, farms, and towns. If problems arose, there was vacant land waiting to be moved to. But when the expansion was done, problems became more visible and less avoidable. People could see that there were "trade-offs" involved in human activity: more industry meant more jobs and more wealth, but there was a price in air and water pollution and human health (among other things).

Things changed, bit by bit. The numbers crept up. More people, more cattle, more problems. And not just in the U.S.A. Nowhere, perhaps, was the price more obvious than in Eastern Europe. The former Soviet Union was infamous for refusing to admit that industrial activity was anything but desirable. Anyone who spoke up about environmental problems risked being shipped off to a gulag in Siberia. The result, which became visible to Western nations after the fall of the Iron Curtain in 1990, was industrial zones where rivers had no fish, children were sickly, and life expectancies were reduced.

By the 1950s and 1960s, Grandpa was looking like an idiot. So were people who said "Grandpa did it. Why can't I?"

The rest of us knew an "Oops" moment when we saw it. We learned why we should not keep doing things the old way. We passed laws such as the Clean Air and Clean Water Acts to force the idiots to change their ways.

But... Lawyers. They invented the concept of "regulatory takings."[4] The idea is that since the Constitution prohibits government from seizing property without due process or compensation, and since environmental regulations prevent certain uses of property and/or increase costs, they amount to just such seizures or takings (of value, if not of actual property). Conservatives use this view of environmental regulations to justify getting rid of those regulations or—sometimes—putting foxes—er, industry lobbyists—in charge of the henhouse—er, regulatory process. Property rights are sacred! Tradition is sacred! Grandpa did it right! And, of course, the only value that counts is economic value. Nature undeveloped is mere wasteland. Don't worry about endangered species. They just can't cope.[5]

Idiots! Excuse me, but... Idiots!

Change happens. Grandpa made mistakes. We learn—or should learn—from those mistakes. We stop doing things the way Grandpa did. We figure out new ways to meet human needs, and—of course!—we make new mistakes. And we learn from them.

Unfortunately, many people want to keep doing things the way Grandpa did. After all, that's how he got rich and built a country and so on. They'd rather ignore the mistakes and the need to learn, the need to change.

That need to change is not about to go away. There are still environmental problems we have to deal with. We also need to consider that science and technology continue to evolve, creating new industries, solving some problems, creating more, and presenting new choices, new opportunities to change. Or not. After all, in due time we will be Grandpa in our turn.

That's what this book is about. I am *not* trying to predict the future. Rather, I am considering a number of areas where we face important choices or opportunities. Which choices we make will have important effects upon our future well-being. The need for some of those choices has been apparent for many years. One prime example involves the choice between a society based on unending growth of both population and industry and one based on sustainability. The former can only succeed for a short time, as was pointed out in the mid-nineteenth century by George Perkins Marsh, after which Mother Nature steps in.

Why have we pursued the route of endless growth? The answers are straightforward: People live better lives—better health, better food, better mobility, better chances for advancement, better social justice. It is not just a matter of rich industrialists—and the politicians they sponsor—getting richer, though that happens too, and it surely motivates much of the relentless push for industrial and scientific and technological progress. Grandpa had his reasons.

Yet there are limits. Surely the wiser course is to recognize this very basic truth and act in such a way as to avoid a head-on collision with those limits. This has clear application to population, food supply, and carbon emissions and global warming.

Among the other issues I have considered in these essays are the impact of Artificial Intelligence (AI) on jobs and a few of the predictable side-effects on identity and identity politics, the need for Guaranteed Basic Income, and the future of higher education. Aside from its effect on jobs, AI also has implications for the future nature of identity theft. It may also bear on whether we are really real, though many people are very doubtful that we live in a far-future simulation. If we choose to continue to pursue the development of AI, we will face important questions about what we do with our lives, how we meet basic needs, whether we need college degrees, and how we protect ourselves against deception.

At a much more mundane level, there are the contents of our bowels, which actually seem to be very important to our health in many ways. Even just as manure, our poop deserves respect for it could be used much more as a fertilizer. It isn't, partly because of our emotional "YUCK!" response to the very idea of taking it seriously. It would be useful if we could get past that "YUCK!" response, but that decision may be beyond the reach of reason.

We also face questions dealing with new technologies such as genetic engineering, synthetic biology, brain implants, self-driving cars. And more. Are they safe? How safe? Should we accept them, or ban them? Unfortunately, many people seem to think that because these technologies differ greatly from what we have been used to for centuries or millennia, we should have nothing to do with them. Sometimes, as with electric cars, our choices are limited by our past investments in an older technology. To make a change requires writing off an extensive infrastructure dedicated to supplying gasoline to users. Changing to solar or wind power is made difficult by an existing infrastructure of power plants of several kinds. Sometimes our choices are driven by immense inertia, perhaps expressed as "That's *not* the way Grandpa did it!"

Inertia is also a factor in planning for disaster, whether an asteroid impact, a massive volcanic eruption, climate change, or a pandemic. Politicians are not good at looking into the future more than just a very few years. Anything past the next election is over the horizon, out of sight and out of mind. With business executives, the horizon of the future may be the end of the current quarter. Forward-looking executives may actually look ahead as far as the end of their career. Anything more is very rare. And the problem of short-sightedness is aggravated by refusal to listen to those who see further, as when a certain president ignores experts and calls journalism the enemy of the people.

Yet there is reason to be hopeful. The Covid-19 pandemic of 2020 hit everyone hard. It also brought home to a great many people that we are all in this together. To slow the spread of the virus, we closed businesses, schools, and entertainment events. We told people to shelter in place—to stay home, in other words. We learned to practice social distancing and wear masks in public not just to protect ourselves, but also to protect others (just in case we were asymptomatic virus-shedders). There was also, of course, a tide of believers in "me-ism" who protested *en masse* against such measures. As a result, many of them were infected with the virus. Some of them died.

We also learned which workers really mattered. Industrialists and bankers and politicians might think they are essential to civilization, but when it came right down to it, the "essential workers" were truckers, delivery people, grocery clerks, and of course the doctors, nurses, EMTs, and other health care workers who labored to keep us alive when the virus struck us down, even at the cost of their own lives.

This book is in large part about choices. If the sense that we are all in this together sticks, we may make much better choices in the future.

ABOUT THE AUTHOR

I've written short stories, novels, textbooks, poetry, and magazine articles for half a century. I spent thirty years as the book columnist for the science fiction magazine *Analog*. With a doctorate in theoretical biology, I also spent more than thirty years as a college professor.

And it's all of a piece. I started reading science fiction in 1957. The appeal to a callow thirteen-year-old was downright traditional, for that was about the age when an awful lot of science fiction fans and writers got hooked. But there was something more in my case. Dad was a biologist, Mom a nurse, so science was in the family. I turned thirteen in 1957, the year of Sputnik. The newspapers were full of Rockets! Satellites! Space! Very soon, we had Mercury, Gemini, and Apollo. Wernher von Braun and Willy Ley were beating the drums for space stations and moon bases.

I was ready to discover *Astounding* (now *Analog)*, with covers that looked (some of the time) just like *Time, Look, Life*, and other magazines of the time. The stories pushed the headlines into the future, and the stories, the fact articles, and even John W. Campbell, Jr.'s editorials (despite Campbell's tendency to chase some ideas into silliness) got me used to thinking about how science and technology interacted with society. Science was important because it meant understanding the universe. Technology meant using that understanding to do cool things. Both could generate exciting stories. But the real message was that science and technology were important precisely because of how they affected people, both as individuals and *en masse*, as a society, not always in obvious ways (electricity changed the rhythm of life because people could stay up later!). A secondary message—blame Campbell's editorials for this one!—was that in order to deal intelligently with issues of science, technology, and society, one must recognize that people disagree about facts, their meaning, and what to do about them, and one must think outside the box (even at the risk of being thought a bit nuts by people who don't even understand that there *is* a box).

In due time, I earned a doctorate in Theoretical Biology from the University of Chicago. When I started writing in the 1970s, it seemed very natural to write science fiction, some of which dealt with those interactions of science, technology, and society. But only a very small number of people can actually make a living writing science fiction these days. When that penetrated, I became a college professor, teaching biology, including environmental science and ecology, as well as other science courses. It was only a part-time gig at first, while I continued to try to make writing generate a full-time income. In due time, the college asked me to develop a course on philosophy and methods of science, which eventually became a science, technology, and society (STS) course. Starting from my course materials, I edited an issues anthology (*Taking Sides: Controversial Issues in Science, Technology and Society*, McGraw-Hill, now in its 14th edition) and developed an anthology of stories on STS themes (*Gedanken Fictions*, Wildside, 2000). A few of the nineteen issues in the *Taking Sides* book dealt with the environment. It seemed a natural progression when the publisher asked me to take over *Taking Sides: Controversial Environmental Issues* (now in its 17th edition), a somewhat more focused STS book.

The shaping influence of Campbell's *Astounding* is still there. Science and technology and their intersection with society are important. Thinking about them is fun. Taking different angles on them (getting outside the box) is crucial. My personal approach has shifted from just reading (Hey! I was just a kid when I started!) to writing stories and articles to doing textbooks. I've shifted my focus, but the same themes are there. They are also in the essays that make up this volume.

My science fiction stories on how genetic engineering may affect the future are collected in the e-book *The GMO Future*.[1] You can find others in *Love Songs and UFOs*.[2]

END NOTES

AI and Jobs

[1] https://www.actsretirement.org/latest-retirement-news/blog/2017/1/23/5-ways-to-beat-post-retirement-depression/

[2] http://yourdost.com/blog/2017/07/the-effects-of-layoffs.html?q=/blog/2017/07/the-effects-of-layoffs.html&

[3] https://www.politicususa.com/2020/04/21/dan-patrick-economy-more-important-than-living.html

[4] Stuart W. Elliott, https://issues.org/byline/stuart-w-elliott/, "Anticipating a Luddite Revival," *Issues in Science and Technology,* Spring 2014, and "Artificial Intelligence, Robots, and Work: Is This Time Different?" *Issues in Science and Technology,* Fall 2018.

[5] Michael A. Hogg, "Radical Change," *Scientific American*, September 2019.

[6] https://en.wikipedia.org/wiki/Albigensian_Crusade

Maintaining a Consumer Economy

[1] Stuart W. Elliott, https://issues.org/byline/stuart-w-elliott/, "Anticipating a Luddite Revival," *Issues in Science and Technology*, Spring 2014, and "Artificial Intelligence, Robots, and Work: Is This Time Different?" *Issues in Science and Technology*, Fall 2018.

[2] https://time.com/collection/davos-2019/5502592/china-social-credit-score/

[3] https://www.thebalance.com/universal-basic-income-4160668

[4] https://www.reuters.com/article/us-health-coronavirus-usa-pelosi/house-speaker-pelosi-says-may-look-at-guaranteed-income-other-aid-idUSKCN2291TO

[5] http://www.canarianweekly.com/minimum-income-spain-launched/

[6] https://www.usnews.com/news/national-news/articles/2020-04-12/pope-ponders-universal-basic-income-as-boris-johnson-is-discharged-opec-strikes-deal

[7]

https://en.wikipedia.org/wiki/Civilian_Conservation_Corps

[8]

https://en.wikipedia.org/wiki/Works_Progress_Administration

[9] Or regulate it. See https://issues.org/perspective-should-artificial-intelligence-be-regulated/.

[10] https://www.stopkillerrobots.org/

[11] "Teamsters Try to Nix UPS Drones, Self-Driving Vehicles," *Automotive Fleet*, January 24, 2018.

Who Needs Education?

[1] https://hechingerreport.org/liberal-arts-face-uncertain-future-at-nations-universities/

[2] Joseph Epstein, "Who Killed the Liberal Arts," *Weekly Standard*, September 17, 2012; https://www.weeklystandard.com/joseph-epstein/who-killed-the-liberal-arts

[3] Stuart W. Elliott, https://issues.org/byline/stuart-w-elliott/, "Anticipating a Luddite Revival," *Issues in Science and Technology*, Spring 2014, and "Artificial Intelligence, Robots, and Work: Is This Time Different?" *Issues in Science and Technology*, Fall 2018.

4 https://www.vice.com/en_us/article/pa9myg/the-hot-new-gen-z-trend-is-skipping-college-v25n3

5 Abigail Hess, "Harvard Business School Professor: Half of American Colleges Will Be Bankrupt in 10 to 15 Years," *CNBC Make It*, November 15, 2017.

6 Derek Thompson, "This Is the Way the College 'Bubble' Ends," *The Atlantic*, July 26, 2017; https://www.theatlantic.com/business/archive/2017/07/college-bubble-ends/534915/.

7 https://thehill.com/opinion/white-house/361674-bread-and-circuses-could-strike-down-america-like-rome

AIdentity Theft

1 https://www.slashgear.com/nvidia-powered-robot-ai-learns-by-watching-humans-20531321/

2 https://www.digitaltrends.com/cool-tech/nvidia-virtual-city-tech/

3 https://www.newscientist.com/article/2143498-deepmind-ai-teaches-itself-about-the-world-by-watching-videos/

4 https://bair.berkeley.edu/blog/2018/10/09/sfv/

5 https://futurism.com/an-artificial-intelligence-program-watched-600-hours-of-youtube-videos-to-study-human-interaction

6 https://www.wired.co.uk/article/internet-of-things-what-is-explained-iot

7 https://www.mindful.org/what-is-mindfulness/

8 https://www.fitbit.com/home

9 https://boingboing.net/2019/01/29/fiat-lux.html

10 https://www.newsweek.com/baby-monitor-hacked-houston-couples-child-threatened-kidnapping-other-lewd-1265060

11 https://www.iotforall.com/5-worst-iot-hacking-vulnerabilities/

12 https://www.technologyreview.com/s/612874/the-

real-reason-america-is-scared-of-huawei-
internet-connected-everything/
[13] https://www.vox.com/the-
goods/2018/11/12/18089090/amazon-echo-
alexa-smart-speaker-privacy-data
[14] https://www.nytimes.com/2018/11/13/us/barriss-
swatting-wichita.html

Are We AIs?

[1] Nick Bostrom, "Are You Living in a Computer
Simulation?" *Philosophical Quarterly*, vol. 53,
No. 211, 2003; https://www.simulation-
argument.com/simulation.pdf.
[2] Joshua Rothman, "What Are the Odds We Are Living
in a Computer Simulation?" *New Yorker*, June
9, 2016
[3] Ricardo Manzotti and Andrew Smart, "Elon Musk Is
Wrong: We Aren't Living in a Simulation,"
Motherboard, June 20, 2016;
https://motherboard.vice.com/en_us/article/y
p3b7w/we-dont-live-in-a-simulation.
[4]

https://www.bleepingcomputer.com/editorial/t
echnology/is-conscious-ai-achievable-and-how-
soon-might-we-expect-it/
[5] https://www.technologyreview.com/s/612913/a-
philosopher-argues-that-an-ai-can-never-be-an-
artist/?utm_campaign=the_download.unpaid.en
gagement&utm_source=hs_email&utm_medium
=email&utm_content=70127287&_hsenc=p2AN
qtz---
PAGFPPy8OgXQzGpkE6UePUlZtVz5Fsi0jTi8h2h
AorScMdpuFyRlxGBcC3aa70hdNIiUOALYAYwn
UcURN7H63tI2FA&_hsmi=70127287
[6] https://futurism.com/artificial-intelligence-hype
[7] Sarah Lewin, "Is the Universe a Simulation?
Scientists Debate," Space.com, April 12, 2016.

[8] Scientists actually wonder whether reality—or
spacetime—is pixelish, though they call it
granularity. It's a matter of quantization. See
https://www.sciencedaily.com/releases/2016/
04/160422115329.htm.

[9] https://www.tcg.vet.cam.ac.uk/about/DFTD

[10] https://plato.stanford.edu/entries/qm-manyworlds/

[11] This reminds me of Terry Pratchett's *Wee Free Men*,
whose title characters, the Nac Mac Feegle,
believe not only that they live in the afterlife,
but also that when they die, they return to the
living world.

Dirt Needs Dirt

[1] When great-grammy died, things changed in a hurry.

[2] For a thorough discussion, see
https://extension.psu.edu/what-is-sewage-
sludge-and-what-can-be-done-with-it.

[3] https://earther.gizmodo.com/the-mad-science-plan-
to-power-the-world-with-poop-
1833062048?utm_campaign=the_download.unp
aid.engagement&utm_source=hs_email&utm_m
edium=email&utm_content=70598668&_hsenc=
p2ANqtz-
9RQvY8Tq2ur1gL2A7RbsTuEgQ0UzgIFG03uBH
YfnA8btJfv4JAP20wyvTc5L9-
Z2wE9U0vMrIPXd_GX3KKaSUmPdfCvg&_hsmi=
70598668

[4] E.g., https://loudounnow.com/2018/10/05/farm-
neighbors-renew-objections-to-use-of-biosolid-
fertilizer/

[5] To be fair, the bacteria, viruses, and parasites to be
found in human manure are those that infect
human beings. (So are some of those in animal
manure.) Coming into contact with sludge can
pose problems. But these problems can be
minimized by composting and prompt plowing.

[6] All by itself, urine is rich in such plant nutrients, and collecting urine has been suggested as a way of avoiding a phosphate crisis. See https://www.buildinggreen.com/news-article/urine-collection-beats-composting-toilets-nutrient-recycling.

[7] https://www.ewg.org/research/dumping-sewage-sludge-organic-farms

[8]

https://www.sciencedaily.com/releases/2016/03/160304092230.htm. See also https://iopscience.iop.org/article/10.1088/1748-9326/ab0071/meta

[9]

https://www.sciencedaily.com/releases/2015/02/150213112142.htm

[10] https://newrepublic.com/article/115883/drugs-drinking-water-new-epa-study-finds-more-we-knew

[11] https://www.smithsonianmag.com/smart-news/sewage-water-reveals-communitys-illegal-drug-habits-180948197/

[12]

https://motherboard.vice.com/en_us/article/z43kk9/the-plan-to-test-cities-sewage-for-drugs-is-a-new-form-of-mass-surveillance-5886b75df1bddf454bd80da8

[13] https://www.scientificamerican.com/article/china-expands-surveillance-of-sewage-to-police-illegal-drug-use/

[14] https://www.wateronline.com/doc/researchers-sequence-sewage-samples-0001

[15] https://cnre.mit.edu/research/development-pipe-crawling-robot-water-pipeline-network-discovery

[16] Might make an interesting CSI show, eh?

[17] https://qz.com/1353825/a-major-us-city-will-start-drinking-its-own-sewage-others-need-to-follow/

[18] https://alumni.berkeley.edu/california-

magazine/just-in/2015-07-06/effluent-
communities-why-drought-will-mean-learning-
drinking
[19] https://www.news24.com/SouthAfrica/News/you-
could-be-drinking-your-neighbours-loo-water-
city-of-cape-town-may-take-water-saving-to-
another-level-20180307
[20] http://theconversation.com/more-of-us-are-
drinking-recycled-sewage-water-than-most-
people-realise-92420

The Frenemy Within

[1]

http://blogs.discovermagazine.com/notrocketsc
ience/2012/08/31/everything-you-never-
wanted-to-know-about-the-mites-that-eat-
crawl-and-have-sex-on-your-
face/#.XF2Ef82neUk
[2]

http://www.mshistorynow.mdah.ms.gov/article
s/241/hookworm-disease-in-mississippi%3A-
the-importance-of-wearing-shoes
[3] https://www.pbs.org/wgbh/nova/article/how-a-
worm-gave-the-south-a-bad-name/
[4]

https://www.ncbi.nlm.nih.gov/pmc/articles/P
MC2841828/
[5] https://medicalxpress.com/news/2015-08-scientists-
benefits-worms.html
[6] https://blog.neogen.com/farm-animals-help-prevent-
asthma-and-allergies-in-kids-study-finds/
[7] http://givemeboldness.blogspot.com/2007/11/peck-
of-dirt.html
[8] https://www.rd.com/health/wellness/triclosan-
antibacterial-soap-dangerous/
[9] https://www.drweil.com/diet-nutrition/food-
safety/best-cutting-board/

---

[10]
  https://www.cdc.gov/parasites/schistosomiasis/biology.html

[11] https://www.who.int/dracunculiasis/disease/en/

[12] https://www.planetdeadly.com/animals/candiru-fish-swim-penis

[13] https://medicalxpress.com/news/2018-08-dog-saliva-bacterial-infection-amputations.html

[14] https://www.popsci.com/brain-eating-amoeba- ⟩ spreading

[15] https://en.wikipedia.org/wiki/Mary_Mallon

[16] https://www.recalls.gov/food.html

Your Other Half

[1] https://www.webmd.com/digestive-disorders/news/20151209/diy-fecal-transplant#1. Don't let the kids read this article—if they do, "playing doctor" could take some very weird turns.

[2] In Massachusetts, OpenBiome (http://www.openbiome.org) collects and screens stool samples from healthy, thin volunteers for distribution to doctors.

[3] https://www.webmd.com/digestive-disorders/news/20190614/fda--infections-1-death-after-fecal-transplants#1

[4]
  https://www.fastcompany.com/2681229/repoopulate-how-fake-poop-can-cure-patients-stomach-ailments

[5]
  https://www.sciencedirect.com/science/article/pii/S2352939318300162

[6]
  https://www.sciencedaily.com/releases/2018/07/180718113343.htm

[7] See Elizabeth Pennisi, "Gut Bacteria Linked to Mental

Well-Being and Depression," *Science*, February 8, 2019.

8

https://www.ncbi.nlm.nih.gov/pmc/articles/PMC5354081/

[9] Does it have to be *your* other half? So far the research has been concentrating on within-species fecal transplants. No one is giving human patients fecal bacteria from dogs or cats. Yet in folk medicine there is such a thing as "nanny-plum tea." This is a tea made from the "plums" (turds) of sheep or goats. It was used to treat many ailments but particularly diarrhea. Since many people believe there is at least some truth to folk medicine, cross-species fecal transplants may warrant some research.

Making It Up

[1] https://humanityplus.org/
[2] Gretchen Vogel, "Embryo Engineering Alarm," *Science*, March 20, 2015.
[3] Jon Cohen, "What Now for Human Genome Editing?" *Science*, December 7, 2018.

4

https://www.sciencemag.org/news/2019/12/chinese-scientist-who-produced-genetically-altered-babies-sentenced-3-years-jail
[5] James Mitchell Crow, "Life 2.0: Inside the Synthetic Biology Revolution," *Cosmos*, April 17, 2018.
[6] https://www.the-scientist.com/news-opinion/dnas-coding-power-doubled-65499
[7] "Vatican Greets First Synthetic Cell with Caution," *America*, June 7-14, 2010.
[8] https://www.wired.co.uk/article/artificial-life-vint-cerf
[9] Eleonore Pauwels, "Public Understanding of Synthetic Biology," *BioScience*, February 2013

---

10 Marion Eisenhut and Andreas P. M. Weber, "Improving Crop Yield," *Science*, January 4, 2019.

It's Loose!

1 http://www.eliminatedengue.com/program
2 For three links to the topic, try these:
https://bigthink.com/stephen-johnson/scientists-are-trying-revive-woolly-mammoth-dna-to-fight-climate-change
https://reviverestore.org/projects/woolly-mammoth/
http://science.sciencemag.org/content/308/5723/796.1
3 https://www.newdinosaurs.com/dire-wolf/
4 http://www.direwolfproject.com/
5 https://reviverestore.org/about-the-passenger-pigeon/
6 https://www.newdinosaurs.com/terror-bird/
7 http://www.fossil-treasures-of-florida.com/terror-bird.html

Making Us Better

1 https://geneticliteracyproject.org/2017/01/16/three-new-disease-resistant-gmos-address-climate-change-save-farmers-billions/
2 https://www.abc.net.au/news/2019-02-25/gene-editing-scientist-may-have-made-the-twins-smarter/10845220
3 https://www.nationalbreastcancer.org/what-is-brca
4 https://futurism.com/genetic-engineering-could-allow-us-to-treat-or-even-prevent-obesity
5

https://motherboard.vice.com/en_us/article/3dm7m9/quadriplegic-man-regains-use-of-arms-with-brain-computer-interface

[6] https://www.scientificamerican.com/article/with-brain-implants-scientists-aim-to-translate-thoughts-into-speech/

[7] https://futurism.com/4-theres-a-brain-implant-that-could-restore-vision-to-the-blind

[8] https://www.hearinglink.org/your-hearing/implants/auditory-brainstem-implants/

[9] https://www.nbcnews.com/mach/science/memory-boosting-brain-implants-are-works-would-you-get-one-ncna868476

[10] https://www.washingtonpost.com/news/the-switch/wp/2016/08/15/putting-a-computer-in-your-brain-is-no-longer-science-fiction/?utm_term=.85222968b924

[11]

https://www.theguardian.com/commentisfree/2018/dec/23/elon-musk-neuralink-chip-brain-implants-humanity

[12] Not that there's anything wrong with being a chimp. Just ask one, once you teach it American Sign Language.

[13] "Transhumanism," *Foreign Policy,* September/October 2004.

[14] "Biomedical Enhancements: Entering a New Era," *Issues in Science and Technology,* Spring 2009. See also Mehlman's book, *Transhumanist Dreams and Dystopian Nightmares: The Promise and Peril of Genetic Engineering* (Johns Hopkins University Press, 2012).

[15] https://emilypost.com/

[16] FaceBook has made us used to that already.

[17] https://www.sciencealert.com/a-talented-bar-owner-in-the-uk-has-built-a-faraday-cage-to-stop-customers-using-their-phones

[18] https://www.amazon.com/Electro-Deflecto-Unisex-Foil-Size/dp/B01I497JAM

[19] http://www.kurzweilai.net/daily-mail-well-be-

uploading-our-entire-minds-to-computers-by-2045-and-our-bodies-will-be-replaced-by-machines-within-90-years-google-expert-claims

[20] https://www.technologyreview.com/s/610456/a-startup-is-pitching-a-mind-uploading-service-that-is-100-percent-fatal/

[21] I still have trouble with streaming. You are *not* supposed to be able to pause a "live" TV show!

How Safe Is Safe?

[1] https://www.darpa.mil/news-events/2014-03-13

[2] https://www.technologyreview.com/the-download/613064/self-driving-cars-are-coming-but-accidents-may-not-be-evenly-distributed/

[3] Nick Kurczewski, "Consumer Confidence in Self-Driving Technology Is Increasing, AAA Study Finds," *Consumer Reports*, January 24, 2018.

[4] Matt McFarland, "The Backlash against Self-Driving Cars Officially Begins," *CNN Money*, January 10, 2017.

[5] M. R. O'Connor, "The Fight for the Right to Drive," *New Yorker*, April 30, 2019.

Are There Limits?

[1] https://www.cnn.com/2018/11/23/health/climate-change-report-bn/index.html

[2] https://www.unenvironment.org/resources/global-environment-outlook-6

[3] David Spratt and Ian Dunlop, "Existential Climate-Related Security Risk: A Scenario Approach," Breakthrough--National Centre for Climate Restoration, May 2019.

[4]

https://www.npr.org/2018/12/03/672893695/david-attenborough-warns-of-collapse-of-civilizations-at-u-n-climate-meeting

[5] https://www.footprintnetwork.org/our-work/ecological-footprint/

[6] https://www.sciencealert.com/we-just-used-up-all-of-earth-s-resources-for-the-year-and-it-s-only-jul

[7] Warren Cornwall, "Even 50-Year-Old Climate Models Correctly Predicted Global Warming," *Science,* December 4, 2019.

[8] https://www.degrowth.info/en/what-is-degrowth/

9 https://edition.cnn.com/world/live-news/coronavirus-pandemic-04-08-20/h_fa578270a745eb4e5b1321742a5a87f4?fbclid=IwAR1ANtYRxMiTyBnR3PemtCwqPSAwUGDLEFN5umaV2RsSVHZ00zT9MvtwCT4

Limiting Factors

1

https://www.theatlantic.com/business/archive/2012/12/33-slides-laying-out-the-crisis-that-could-lead-to-mass-world-starvation/265909/

[2] Try these links for the straight dope:
https://www.groundwater.org/get-informed/groundwater/overuse.html
https://water.usgs.gov/edu/gwdepletion.html
https://news.nationalgeographic.com/news/2014/08/140819-groundwater-california-drought-aquifers-hidden-crisis/
https://www.usgs.gov/news/usgs-high-plains-aquifer-groundwater-levels-continue-decline
https://phys.org/news/2018-08-farmers-groundwater-giant-ogallala-aquifer.html

[3] David Pimentel, et al., "Will Limited Land, Water, and Energy Control Human Population Numbers in the Future?" *Human Ecology,* August 2010.

[4] https://www.cnn.com/2019/08/06/world/aqueduct-

water-climate-crisis-intl-scli/index.html
[5] Avery Thompson, "Cape Town Wants to Tow an Iceberg to Solve Its Water Shortage," *Popular Mechanics*, July 6, 2018; https://www.popularmechanics.com/science/environment/a22064173/cape-town-wants-to-tow-an-iceberg-to-solve-its-water-shortage/
[6] It is also a Biblical concept. Recall Pharoah's dream of seven fat kine and seven lean kine?

What's for Dinner?

[1] https://www.census.gov/popclock/
[2] https://www.nbcnews.com/better/lifestyle/can-your-diet-save-planet-ncna1044126
[3] https://impossiblefoods.com/food/
[4] https://www.scientificamerican.com/article/lab-grown-meat/
[5]

https://news.stanford.edu/news/2006/november8/ocean-110806.html
[6] Eva Plaganyi, "Climate Change Impacts on Fisheries," *Science*, March 1, 2019.
[7]

https://www.nytimes.com/2019/07/10/business/fake-fish-impossible-foods.html
[8] https://www.iflscience.com/environment/will-we-all-be-eating-insects-50-years/
[9] https://www.scientificamerican.com/article/as-insect-populations-decline-scientists-are-trying-to-understand-why/ ; the latest report—February 2019—indicates the decline of insects is global, rapid, and a serious threat to ecosystems; causes are thought to be climate change, urbanization, and the use of pesticides-

https://www.cnn.com/2019/02/11/health/insect-decline-study-intl/index.html

---

[10] https://www.mnn.com/earth-matters/animals/stories/what-windshield-phenomenon

[11] https://www.npr.org/sections/goatsandsoda/2016/08/10/489302134/should-we-make-room-for-worms-on-our-dinner-plate

[12] https://aggie-horticulture.tamu.edu/archives/parsons/vegetables/colecrop.html

[13] https://geneticliteracyproject.org/2014/07/01/rust-threatening-wheat-crops-worldwide-could-be-thwarted-with-genetics-but-anti-gmo-fervor-remains-challenge/

[14] https://www.acsh.org/news/2017/07/30/genetically-engineered-wheat-reduces-need-fertilizer-helps-environment-11627

Bouncing, Bouncing, Bouncing

[1] Melinda Harm Benson and Robin Kundis Craig, "The End of Sustainability," *Society & Natural Resources,* vol. 27, No. 7, 2014, http://dx.doi.org/10.1080/08941920.2014.901467. A much shorter version of this paper appeared at http://ensia.com/voices/the-end-of-sustainability/.

[2] https://www.cnn.com/2020/04/22/africa/coronavirus-famine-un-warning-intl/index.html?fbclid=IwAR1zjo95znZktZf7FsXuwXBV3gmKuwxSZOEPTh2PhP-fjT31L8sAEl6GQQw

[3] New European climate models say a worst-case scenario may mean the global average temperature goes up as much as 7 degrees Celsius (12 degrees Fahrenheit):

http://www.cnrs.fr/en/two-french-climate-models-consistently-predict-pronounced-global-warming

Gee, It's Lonely Here

[1] There have been more than fifty searches for extraterrestrial (ET) radio signals since 1960. The earliest ones were very limited. Later searches have been more ambitious, using multiple telescopes and powerful computers to scan millions of radio frequencies per second. New technologies and techniques continue to make the search more efficient. See James Shreeve, "Who's Out There?" *National Geographic*, March 2019.

[2] S. G. Wallenhorst, "The Drake Equation Reexamined," *Quarterly Journal of the Royal Astronomical Society*, Vol. 22, p. 380, 1981.

[3] https://www.sciencealert.com/here-s-what-would-happen-if-solar-storm-wiped-out-technology-geomagnetic-carrington-event-coronal-mass-ejection

[4] As a theoretical biologist, I'm optimistic that where life is possible, it will happen.

[5] Frank Drake notes that it takes only one very long-lived civilization to make the *average* lifetime *much* larger. That, of course, would throw off the calculation by overestimating **N**. See "The Drake Equation Revisited, Part I," *Astrobiology Magazine*, September 29, 2003.

The Cassandra Code

[1] https://www.cnn.com/2019/06/04/health/climate-

change-existential-threat-report-intl/index.html

Just A-Looking for a Home

1

https://www.apnews.com/65694195c91d4b62
b275bd14a6955b4c

2 https://www.cnn.com/2019/07/03/asia/india-heat-
wave-survival-hnk-intl/index.html

3

https://www.npr.org/2019/12/21/790444258
/catastrophic-wildfires-continue-to-rage-across-
australia
4 https://www.bbc.com/news/world-asia-india-
49234708
5 https://www.npr.org/2019/08/19/752274659/no-
joke-trump-really-does-want-to-buy-greenland
6 https://www.cnn.com/2019/07/17/asia/india-
nepal-flooding-climate-refugees-intl-
hnk/index.html
7 http://www.young-diplomats.com/who-are-indias-
allies/
8 https://www.cnn.com/2019/08/02/politics/nuclear-
treaty-inf-us-withdraws-russia/index.html
9 https://www.africa-eu-partnership.org/en
10 Or IV. As I write this the US has just assassinated an
Iranian general and the saber-rattling is at its
loudest pitch in years.
11 https://www.energy.gov/eere/solar/solar-
desalination
12 https://ehs.unu.edu/blog/5-facts/5-facts-on-
climate-migrants.html
13 https://www.youtube.com/watch?v=oQOPTEd24l8

The Problem of Sunk Costs

[1] "Inequality, Sunk Costs, and Climate Policy," *Triple Crisis*, February 27, 2019; http://triplecrisis.com/inequality-sunk-costs-and-climate-policy/.

Throwing Shade

[1] https://www.straitstimes.com/world/climate-changing-faster-than-feared-but-why-are-we-surprised

[2] Matt McGrath, "Final Call to Save the World from 'Climate Catastrophe.'" *BBC News*, October 8, 2018.

[3] https://www.theguardian.com/environment/2019/dec/10/greenland-ice-sheet-melting-seven-times-faster-than-in-1990s

[4] It's already happening: https://www.cnn.com/2018/12/04/us/camp-fire-insurance-company-liquidation/index.html

[5] https://www.cnn.com/2019/08/25/americas/amazon-fire-efforts-damage/index.html

[6] https://www.politifact.com/truth-o-meter/statements/2016/jun/03/hillary-clinton/yes-donald-trump-did-call-climate-change-chinese-h/

[7] https://insideclimatenews.org/news/22082017/study-confirms-exxon-misled-public-about-climate-change-authors-say

[8] Download the 2018 Fourth National Climate Assessment at https://nca2018.globalchange.gov/downloads/

[9] https://www.smithsonianmag.com/history/blast-from-the-past-65102374/

[10] "Albedo Enhancement by Stratospheric Sulfur Injections: A Contribution to Resolve a Policy

Dilemma?" *Climate Change*, August 2006,
[11] "Feasibility of Cooling the Earth with a Cloud of Small Spacecraft near the Inner Lagrange Point (L1)," *Proceedings of the National Academy of Sciences of the United States of America*, November 14, 2006.
[12] Tim Smedley, "How Artificially Brightened Clouds Could Stop Climate Change," BBC Future (February 26, 2019).
[13] https://techcrunch.com/2018/10/23/y-combinator-issues-a-request-for-geo-engineering-startups-because-climate-change-is-real-and-were-all-going-to-die/
[14] "Pursuing Geoengineering for Atmospheric Restoration," *Issues in Science and Technology*, Summer 2010.
[15] *Climate Intervention: Carbon Dioxide Removal and Reliable Sequestration* (2015) and *Climate Intervention: Reflecting Sunlight to Cool Earth* (2015).
[16] James Temple, "The Growing Case for Geoengineering," *Technology Review*, April 18, 2017.
[17] Katherine Ellison, "Why Climate Change Skeptics Are Backing Geoengineering," *Wired*, March 28, 2018.
[18] https://www.technologyreview.com/s/614991/the-us-government-will-begin-to-fund-geoengineering-research/
[19] "The Climate Engineers," *Wilson Quarterly*, Spring 2007.
[20] Geoengineering Monitor, "Hands Off Mother Earth! Manifesto Against Geoengineering," October 2018; http://www.geoengineeringmonitor.org/2018/10/hands-off-mother-earth-manifesto-against-geoengineering/.
[21] https://www.unicef-irc.org/article/920-climate-

change-and-intergenerational-justice.html;
http://ww.hettingern.people.cofc.edu/Environm
ental_Philosophy_Sp_09/Gardner_Perfect_Moral
_Storm.pdf

Collapse

1

   https://www.nationalgeographic.com/news/20
   11/3/110302-solar-flares-sun-storms-earth-
danger-carrington-event-science
2 https://www.census.gov/popclock/

Gimme Juice

1 Towns were smaller then, too.
2

   https://www.forbes.com/sites/johnparnell/201
8/12/30/new-chinese-solar-plant-undercuts-
cost-of-coal-power/#7db224ed1182
3 https://e360.yale.edu/digest/renewables-cheaper-
than-75-percent-of-u-s-coal-fleet-report-finds
4 https://thinkprogress.org/renewables-now-cheaper-
than-new-coal-or-gas-across-two-thirds-of-the-
world-c4980412cb53/
5 http://knowledge.wharton.upenn.edu/article/can-
the-world-run-on-renewable-energy/
6 https://www.scientificamerican.com/article/exxon-
knew-about-climate-change-almost-40-years-
ago/
7 http://www.pbs.org/now/shows/347/oil-
politics.html
8 https://www.carbonbrief.org/daily-brief/g7-leaders-
agree-to-phase-out-fossil-fuel-use-by-end-of-
century
9 https://unfccc.int/news/g20-must-phase-out-fossil-
fuel-subsidies-by-2020
10

https://www.npr.org/2018/05/17/612082781
/california-to-require-all-new-homes-to-have-
solar-panels-starting-in-2020

[11]

https://www.csmonitor.com/Environment/201
8/0911/California-to-phase-out-fossil-fuels-by-
2045

[12] https://news.abs-
cbn.com/business/01/27/19/germany-should-
phase-out-coal-use-by-2038-commission

[13]

https://www.greencarreports.com/news/11205
65_a-million-electric-cars-sold-in-the-u-s-led-
by-california

[14] Peter Fairley, "The H2 Solution,"
Scientific American, February 2020.

Let It Glow

[1] http://www.world-nuclear.org/information-
library/safety-and-security/safety-of-
plants/fukushima-accident.aspx

[2] Don't you dare dive in!

[3] The Kyshtym disaster--
https://en.wikipedia.org/wiki/Kyshtym_disaste
r.

[4] https://www.nrc.gov/waste/spent-fuel-storage/dry-
cask-storage.html

[5] E.g., Chris Williams, "The Case against Nuclear
Power," *International Socialist Review,* May
2011.

[6] https://www.energy.gov/ne/nuclear-reactor-
technologies/small-modular-nuclear-reactors

[7] Jessica Lovering and Kenton De Kirby, "Why the
United States Should Partner with Africa to
Deploy Advanced Reactors," *Issues in Science*

and Technology, Winter 2019.

[8] Adrian Cho, "The Little Reactors that Could," Science (February 22, 2019).

[9] https://e360.yale.edu/features/why-nuclear-power-must-be-part-of-the-energy-solution-environmentalists-climate

[10] Timothy A. Frazier, "The Role of Policy in Reviving and Expanding the United States' Global Nuclear Leadership," Center on Global Energy Policy, Columbia University, March 2017.

[11] https://prod-ng.sandia.gov/techlib-noauth/access-control.cgi/2009/094401.pdf

[12] https://www.tri-cityherald.com/news/local/hanford/article152258757.html

Shine On!

[1] It's the same idea as leaving a garden hose on the lawn—the water inside may not get hot enough to make a cup of tea, but it will do for a hot shower.

[2] http://www.htproducts.com/solar-flatpanelcollectors.html

[3] https://www.dezeen.com/2011/06/28/the-solar-sinter-by-markus-kayser/

[4] "Beleaguered California Solar Plant Finally Produces Enough Power," *The American Interest,* March 4, 2017.

[5] Sebastian Anthony, "World's First Solar Road Opens in France: It's Ridiculously Expensive," *Ars Technica,* December 23, 2016.

[6] https://www.tesla.com/solarroof

[7] https://www.npr.org/2018/05/17/612082781/california-to-require-all-new-homes-to-have-solar-panels-starting-in-2020

[8] Peter Kind, "Disruptive Challenges: Financial

Implications and Strategic Responses to a Changing Retail Electric Business" (Edison Electric Institute, 2013).

[9] http://www.nbc-2.com/story/24790572/cape-woman-living-of-the-grid-challenged-by-city#.VGjCsMltgpE

[10] Monica Castaneda, et al., "Myths and Facts of the Utility Death Spiral," *Energy Policy*, November 2017.

[11] It is worth noting that "wind-heavy or solar-heavy U.S.-scale power generation portfolios could in principle provide ~80% of recent total annual U.S. electricity demand." See Matthew R. Shaner, Steven J. Davis, Nathan S. Lewis, and Ken Caldeira, "Geophysical Constraints on the Reliability of Solar and Wind Power in the United States," *Energy & Environmental Science,* February 2018. Since neither wind nor sun are available 24/7, a modernized "smart grid" will be needed to coordinate a host of small and intermittent electricity sources, such as home solar and wind power.

[12] Utilities even argue that coal power is needed for this purpose.

[13] https://ips-dc.org/dirty-energys-quiet-war-on-solar-energy/

[14] https://climatecrocks.com/2018/06/15/in-face-of-trumps-war-on-renewables-solar-makes-big-inroads-on-coal/

Prepping for Disaster

[1] Paula Papale and Warner Marzocchi, "Volcanic Threats to Global Society," *Science*, March 22, 2019.

[2] https://www.cnn.com/2018/12/24/asia/indonesia-tsunami-disaster-year-intl/index.html

[3] https://www.theglobalist.com/the-cost-of-a-human-

life-statistically-speaking/

[4] https://www.washingtonpost.com/news/speaking-of-science/wp/2018/04/20/the-yellowstone-supervolcano-is-a-disaster-waiting-to-happen/?utm_term=.db192ed98e6d

[5] http://www.meteorcrater.com/

[6] Paul Voosen, "Massive Crater under Greenland's Ice Points to Climate-Altering Impact in the Time of Humans," *Science*, November 14, 2018.

[7] https://earthsky.org/space/what-is-the-tunguska-explosion

[8] https://www.cnet.com/news/nasa-head-expect-a-major-asteroid-strike-in-your-lifetime/

[9] https://www.whitehouse.gov/wp-content/uploads/2018/06/National-Near-Earth-Object-Preparedness-Strategy-and-Action-Plan-23-pages-1MB.pdf

[10] https://cneos.jpl.nasa.gov/ca/

[11] https://www.history.com/topics/middle-ages/black-death

[12] https://www.cdc.gov/flu/pandemic-resources/basics/past-pandemics.html

[13]

https://www.sciencemag.org/news/2020/03/modelers-weigh-value-lives-and-lockdown-costs-put-price-covid-19

If It Quacks like a DURC...

[1] Including ducks, of course.

[2] https://www.cdc.gov/features/1918-flu-pandemic/index.html

[3] The potential for accidental releases from labs was also a major concerns; see Laurie Garrett, "The Bioterrorist Next Door," *Foreign Policy,* December 15, 2011, Fred Guterl, "Waiting to Explode," *Scientific American,* June 2012, and

Tina Hesman Saey, "Designer Flu," *Science News,* June 2, 2012.

[4] Jon Cohen, "Does Forewarned = Forearmed with Lab-Made Avian Influenza Strains?" *Science,* February 17, 2012.

[5] Masaki Imai, et al., "Experimental Adaptation of an Influenza H5 HA Confers Respiratory Droplet Transmission to a Reassortant H5 HA/H1N1 Virus in Ferrets" (https://www.nature.com/articles/nature10831), and Sander Herfst, et al., "Airborne Transmission of Influenza A/H5N1 Virus Between Ferrets" (http://www.sciencemag.org/content/336/608 8/1534.full).

[6] David Malakoff and Martin Enserink, "New U.S. Rules Increase Oversight of H5N1 Studies, Other Risky Science," *Science,* March 1, 2013.

[7] Jocelyn Kaiser, "Controversial Flu Studies Can Resume, U.S. Panel Says," *Science,* February 15, 2019.

[8] National Academies of Sciences, Engineering, and Medicine; Policy and Global Affairs; Committee on Science, Technology, and Law; Committee on Dual Use Research of Concern: Options for Future Management, *Dual Use Research of Concern in the Life Sciences: Current Issues and Controversies* (National Academies Press, 2017).

[9] There is also "sensitive but unclassified" or "controlled unclassified information," which can mean restrictions and redactions of key details. See https://fas.org/sgp/cui/index.html.

[10] Gina Bari Kolata, "Cryptography: A New Clash Between Academic Freedom and National Security," *Science*, August 29, 1980.

[11] https://www.snopes.com/fact-check/drano-bottle-bombs/

[12] Phillip A. Sharp, "1918 Flu and Responsible Science"

[Editorial], *Science*, October 7, 2005.

[13] Yudhijit Bhattacharjee, "Should Academics Self-Censor Their Findings on Terrorism?" *Science*, May 19, 2006.

[14] https://www.thoughtco.com/green-revolution-overview-1434948

Dealing Green

[1] https://www.congress.gov/bill/116th-congress/house-resolution/109/text

[2] https://www.foxnews.com/opinion/democrats-green-new-deal-is-a-crazy-new-deal-that-would-be-a-disaster-for-us-all

[3] https://www.gp.org/gnd_full

[4] https://www.sierraclub.org/trade/what-green-new-deal

[5] https://www.thenation.com/article/democrats-green-new-deal/

[6] https://nca2018.globalchange.gov/downloads/

[7]

https://en.wikipedia.org/wiki/Civilian_Conservation_Corps

[8]

https://en.wikipedia.org/wiki/Works_Progress_Administration

[9] The rash of sex abuse scandals engulfing both Catholic and Protestant institutions argues against the idea that churches should have any special status at all.

Do It Now

1 https://www.nbcnews.com/news/world/global-climate-strike-protests-expected-draw-millions-n1056231

[2] https://www.cnn.com/2019/09/23/us/washington-climate-shutdown-trnd/index.html

End Notes

3 https://www.un.org/en/climatechange/

4

    https://www.un.org/en/climatechange/assets/
pdf/CAS_closing_release.pdf

5 http://www.cnrs.fr/en/two-french-climate-models-
consistently-predict-pronounced-global-
warming

6 https://www.cnn.com/2019/09/25/world/un-ipcc-
report-oceans-and-ice-climate-
change/index.html

7 https://www.cnn.com/2018/10/07/world/climate-
change-new-ipcc-report-wxc/index.html

8 https://www.ases.org/removing-carbon-from-the-
atmosphere/

9 https://www.technologyreview.com/s/614382/new-
york-city-has-big-plansand-20-billionto-save-
itself-from-climate-
change/?utm_source=newsletters&utm_mediu
m=email&utm_campaign=the_download.unpaid.
engagement

10 https://time.com/5622374/donald-trump-climate-
change-hoax-event/

11 https://www.huffpost.com/entry/bolsonaro-
un_n_5d8a2989e4b0c2a85cb1ac8c

12 https://www.cnn.com/2019/09/20/us/greta-
thunberg-profile-weir/index.html

13 https://www.scidev.net/global/climate-
change/news/un-gives-12-year-deadline-to-
crush-climate-change.html

Enemy of the People

1 http://www.colby.edu/lovejoyaward/the-story-of-
elijah-parish-lovejoy/

2

    http://www.americaslibrary.gov/jb/reform/jb_r
eform_lovejoy_1.html

3 http://www.colby.edu/lovejoyaward/

Grandpa Did It. Why Can't I?

[1] See e.g. Courtney Sherwood and Kirk Johnson, "Bundy Brothers Acquitted in Takeover of Oregon Wildlife Refuge," *New York Times* (October 27, 2016).

[2] https://www.justice.gov/enrd/multiple-use-lands

[3] Okay, Great-grandpa. Maybe even Great-great-grandpa.

[4] See William A. Fischel, *Regulatory Takings: Law, Economics, and Politics* (Harvard University Press, 1995).

[5] To borrow a conservative line about poor people, maybe endangered species just made bad life choices.

About the Author

[1] https://www.amazon.com/Thomas-Eastons-GMO-Future-MEGAPACK-ebook/dp/B01M687KMG

[2] https://www.amazon.com/Thomas-Eastons-Love-Songs-MEGAPACK%C2%AE-ebook/dp/B01N3NR1Z4

www.ingramcontent.com/pod-product-compliance
Lightning Source LLC
Chambersburg PA
CBHW022037190326
41520CB00008B/608